The Institute of Biology's
Studies in Biology no. 75

Viviparity

Peter J. Hogarth
B.Sc., D.Phil.
Department of Biology University of York

Edward Arnold

First published 1976
by Edward Arnold (Publishers) Limited
25 Hill Street, London W1X 8LL

Boards edition ISBN: 0 7131 2592 6
Paper edition ISBN: 0 7131 2593 4

Printed in Great Britain by
The Camelot Press Ltd, Southampton

General Preface to the Series

It is no longer possible for one textbook to cover the whole field of Biology and to remain sufficiently up-to-date. At the same time teachers and students at school, college or university need to keep abreast of recent trends and know where the most significant developments are taking place.

To meet the need for this progressive approach the Institute of Biology has for some years sponsored this series of booklets dealing with subjects specially selected by a panel of editors. The enthusiastic acceptance of the series by teachers and students at school, college and university shows the usefulness of the books in providing a clear and up-to-date coverage of topics, particularly in areas of research and changing views.

Among features of the series are the attention given to methods, the inclusion of a selected list of books for further reading and, wherever possible, suggestions for practical work.

Readers' comments will be welcomed by the author or the Education Officer of the Institute.

1976

The Institute of Biology,
41 Queens Gate,
London, SW7 5HU

Preface

Viviparity is generally thought of as the prerogative of mammals. Like many generalizations, this is false. Some mammals lay eggs, while many non-mammals give birth to live young. Representatives of practically every class of the vertebrates, and of many groups of invertebrates, reproduce in this way. Indeed, the sporadic distribution of viviparity in the animal kingdom bears little relation to the evolutionary line which produced the mammals.

It is not always easy to avoid the chauvinism of considering other animals from an exclusively mammalian viewpoint. Due to our relative ignorance of the physiology of animals other than mammals, many of the concepts available in discussing viviparity in the lower vertebrates come from work on mammals. I hope this will not obscure the fact that viviparity can only be understood by considering it as an adaptation evolved in diverse situations as a solution to diverse problems; viviparous animals are interesting in their own right, not for what they tell us about our own ancestry.

York, 1976

P. J. H.

Contents

1 Introduction

1.1 What is viviparity?

The distinction between oviparity—the laying of eggs—and viviparity—the production of live young—appears at first sight to be a straightforward one. For example, birds clearly fall into the former category and mammals into the latter. In the course of evolution, viviparous species presumably evolved from oviparous ancestors by the interpolation of a progressively longer interval between fertilization and laying of the eggs. In the ancestry of a living viviparous species, there are intermediate forms which are hard to assign to either category. Likewise, present-day animal species have a diverse spectrum of reproductive methods which it is not always easy to classify.

In this book I propose to use the term 'viviparous' to include all species in which the eggs are fertilized internally and are retained within the body of the mother for a significant time before deposition.

Some authorities have created the term 'ovoviviparity' for animals which retain their fertilized eggs, but where the eggs are not physiologically dependent on the mother. In this situation, no nutrients pass from the mother to the embryo and the situation is more a form of internal incubation than of pregnancy. An example is the guppy.

However, since we are largely ignorant of the metabolic relationships between the mother and the embryo, the distinction between viviparity and ovoviviparity cannot be made with any confidence.

The retention of the egg membrane during gestation has been used as a criterion of ovoviviparous reproduction, but it has been shown that physiological and anatomical adaptation need not necessarily be related. For instance, in some rays the embryo obtains nutritional support from the mother despite the fact that the egg membrane is retained throughout development. However, the embryos of certain sharks which have a complex placental structure are considerably less dependent on the mother. This means that when we lack information about the physiological aspects of reproduction, the anatomical evidence is no substitute.

Thus, in the present state of knowledge, I do not consider ovoviviparity to be a useful concept, and I shall use the term viviparous to include all the types described by other authors as ovoviviparous, as well as those conventionally classified as viviparous.

1.2 Strategies for reproduction

The survival of any species depends on it having a method of reproduction which ensures the production of enough offspring to allow for mortality due to predation, disease and accidental death and to maintain the population at a constant level. In addition, some degree of dispersal of the species is advantageous to maintain genetic diversity and to reduce the impact of local catastrophes. Various reproductive strategies have evolved to achieve these basic requirements for a successful species.

Contrasting reproductive strategies are displayed by the cod, where the female produces several million eggs, and the guppy, a viviparous species which produces only 20 or 30 live young in each brood. Superficially, viviparity appears to be more efficient and desirable; the wastage of the vast majority of the eggs is avoided and the resulting economy of effort is taken in itself to be sufficient advantage to explain why viviparity evolved. What such a view does not explain is how the cod manages to survive if it is so inefficient. In fact, the comparison is totally inappropriate and the conclusion fallacious because both cod and guppy are successful species. In each case, each pair of fish produces on average two surviving offspring.

As far as dispersal of the young is concerned, this is ensured in the cod by letting the eggs and larvae float passively in the sea. A necessary corollary of this is the production of very large numbers of eggs. Each egg has rather little yolk and hatches into a larva which feeds on plankton.

The broadcasting of numerous small eggs would not be appropriate to the guppy's ecological situation. Guppies live in small freshwater pools and watercourses. Production of millions of eggs would not result in wider dispersal than could be achieved by a pregnant female swimming until she finds a suitable place to give birth; freshwater plankton is generally too sparse to support planktonic fish larvae; and since fresh water is less dense than sea water, it is much less likely that a floating egg could evolve.

Viviparity as a reproductive mechanism can evolve in the sea, but not in a species already adapted to produce millions of eggs. In most cases, it seems likely that viviparity has evolved in species which already produced a rather small number of eggs, each with sufficient nutrient for its full development. Retention of these eggs after fertilization would thus require only slight physiological adaptation, and the transition might result from correspondingly slight selective advantage.

Many advantages might accrue from viviparity. Carrying eggs for a time after fertilization may give the female a better opportunity to select a site for the deposition of her offspring. Retention of the young might also serve as a means of protecting them against predators. The adult may be better able to withstand the rigours of a physiologically demanding environment than the young. Possible examples of such environments are aquatic habitats low in oxygen, or low in salt, as estuarine and fresh water is less dense than sea water, it is much less likely that a floating egg removes the risk of desiccation of eggs or young.

Specific situations in which such advantages of viviparous reproduction may be important will be discussed in later sections of this book, particularly in the last chapter. It is worth pointing out, however, that in only a very few cases have species been sufficiently investigated to say with any assurance what advantages are brought by viviparity. It is relatively easy to ascribe benefits to viviparity in a particular case, but practically impossible to say why viviparity should have evolved rather than other, possibly equally beneficial alternatives.

Possible modes of reproduction, then, will be limited by the physiological heritage of the species in question, and the advantage of any given method will depend on the ecology of that species. Different methods of reproduction may be regarded as strategies for survival of the species and no one mode of reproduction is consistently superior to others. Different circumstances make different modes more or less valuable and effective.

1.3 Problems of pregnancy

In a truly viviparous animal, the eggs must be fertilized within the body of the mother (although as we shall see later, some of the more eccentric forms of viviparity are exceptions to this). Viviparity could have evolved only in a species which had already adopted internal fertilization. The problem of ensuring internal fertilization is common to all groups with viviparous members and has been solved by the evolution of a variety of intromittent organs equivalent to the penis of mammals. Penis analogues do not, however, have a common origin and ancestry and as we shall see they have evolved from a number of different structures. This is evidence that viviparity itself must have evolved independently numerous times.

A large number of animals with internal fertilization show some form of viviparity. This suggests that, to a species which had achieved a method of internal fertilization, the further development of viviparity represented only a relatively slight transition. Presumably this condition originated as a progressively longer delay between fertilization and the deposition of the fertilized eggs. The endocrine control of these two events must therefore have become separated; parturition eventually being elicited by a different stimulus from that which controlled ovulation and mating behaviour. The problem of endocrine adaptation to the retention of fertilized eggs must be solved before true viviparity is possible.

Once viviparity is established, various further physiological adaptations may evolve. The female must be able to supply whatever the developing embryo would otherwise have obtained from its physical surroundings. These requirements, for example, the supply of oxygen and water and the removal of carbon dioxide and other waste materials, probably necessitate little adaptation on the mother's part. The supply of nutrients to the embryo is physiologically more demanding and may be associated with complex placental structures and further endocrine adaptations by the mother.

Finally, the fetus represents foreign material within the mother. Half its characteristics derive genetically from the father, and it is thus in a position very similar to that of a skin graft. Such a skin graft (at least in vertebrates) would rapidly be rejected in an immune response: why does this not happen to the fetus?

Solutions to these problems are a necessary concomitant to the evolution of viviparity. The problems are common to all viviparous animals, but the solutions are different. It is the variety of the solutions which supplies much of the interest of the study of viviparity.

2 Sharks and Rays

2.1 The environment of the embryo

In the elasmobranchs (sharks, dogfish, skates and rays), the osmotic concentration of the blood is close to that of the surrounding sea water. This is very different from the situation in most other fish (Fig. 2–1), and is achieved by retaining urea [$CO(NH_2)_2$] within the body instead of eliminating it from the gills and kidney as a waste product. The high level of urea in the blood of elasmobranchs (uraemia) would prove fatal to any other fish; elasmobranch tissues not only survive a high urea level, they cannot function normally without it.

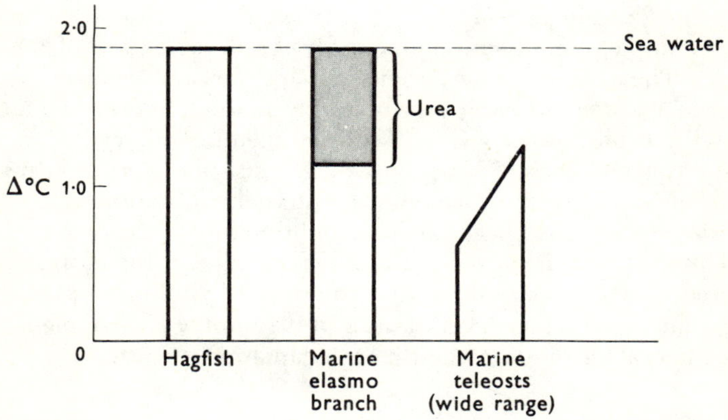

Fig. 2–1 Osmotic concentration in the blood of hagfish, marine elasmobranchs and marine teleosts. The high concentration of urea in the elasmobranch blood renders it isosmotic. Δ°C=freezing-point depression, and is proportional to osmotic concentration; Δ for sea water=1.86°. (After Lockwood, A. P. M. (1963), *Animal Body Fluids and their Regulation*. Heinemann Educational, London.)

This is true of embryos as well as adults. Initially however, the embryonic kidneys are incapable of retaining urea. In egg-laying elasmobranchs (for instance, skates) the concentration of urea necessary for embryonic survival is maintained by an egg-case which is impermeable to urea, as well as giving physical protection. This egg-case is the 'mermaid's purse', often found on British shores. As the embryonic kidney matures and develops the ability to retain urea, mucus plugs in the egg-case dissolve and allow sea water to circulate round the embryo.

The egg-case is complete before the egg is laid. Because spermatozoa cannot penetrate it, fertilization must also occur before laying. Sperm are deposited in the oviduct by the 'claspers' of the male, pelvic fins modified to serve the same function as the penis of mammals. Fertilization occurs in the upper oviduct and the developing embryo passes down the oviduct to the exterior.

In the viviparous spiny dogfish (*Squalus acanthias*), the embryos are delayed in their passage to the exterior, and retained in the oviduct for a gestation period of 18 to 22 months. The lower part of the oviduct is modified to form a uterus. Until the sixth month of gestation, the embryonic kidneys are immature, and the urea concentration in the fluids of the uterus is maintained by the mother at the same level as in her blood. After six months, the uterus is periodically flushed with sea water from the exterior, and the urea concentration is therefore very low, but the embryo is now able to maintain its own uraemia.

This situation parallels so exactly that of egg-laying elasmobranchs that it is apparent that viviparity in *S. acanthias* serves the same function as does the production of urea-impermeable egg-case, i.e. that of maintaining a suitable environment for the development of the early embryo. The inability of the elasmobranch embryo to regulate urea loss in its early stages probably constituted a major selection pressure towards the evolution of viviparity in the group, while even oviparous elasmobranchs (presumably including the ancestors of modern viviparous species) require a method of internal fertilization; which, as we have already noted is a necessary precondition for viviparity to evolve.

Viviparous and oviparous elasmobranchs appear, therefore, to have achieved different solutions to the common physiological problem of maintaining adequate concentrations of urea in the environment of the fetus; internal fertilization being a pre-adaptation to both.

2.2 Relations with the mother

Retention of the fertilized eggs until they hatch, the most primitive form of viviparity, does not require major morphological adaptations in the female. The reproductive system of viviparous species is generally similar to that of oviparous ones. The oviparous spotted dogfish (*Scyliorhinus canicula*) has a single (right) ovary which releases eggs into the funnel formed by the fused ends of the two oviducts. The eggs pass through the nidamental gland, where they accumulate various nutrients, are fertilized, and are finally encased in the horny egg-case. They then pass along the oviduct to the exterior (Fig. 2–2).

In viviparous species, the egg-case, and the part of the nidamental gland which secretes it, are much reduced; otherwise, the gross anatomy of the genital tract is very similar to that of the egg-laying forms. The only major exception to this generalization is in the Hypotremata (skates and rays). In the oviparous skates (*Raja* spp.), both ovaries are present, while the viviparous rays normally retain only the left ovary and oviduct.

Within this general plan, variations occur. The histological structure of the lining of the uterus differs from species to species; these differences are related to specializations in nutrition of the fetus (see section 2.3, p. 9).

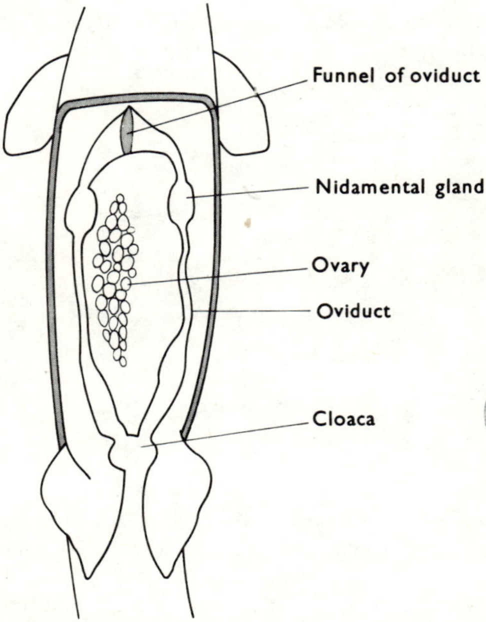

Fig. 2–2 Reproductive system of a female elasmobranch.

With the onset of pregnancy, changes may occur in the uterus. The non-pregnant uterus of the spiny dogfish (*Squalus acanthias*) is a simple tube, lined by a single layer of columnar epithelium, and with a sparse blood supply. During pregnancy, the uterus expands greatly and becomes highly convoluted, increasing its surface area several hundred times. The lining epithelium becomes much thinner, and the capillary network beneath becomes very extensive. These changes probably serve to facilitate exchange of gases and other materials between mother and fetus.

In other species, particularly in the viviparous rays, the area of the uterus lining is increased by the growth of many small flaps of tissue. In an immature female of the round stingray (*Urolophus halleri*), these flaps are 2–3 mm in length; shortly before ovulation they measure 5 mm. If fertilization occurs and an embryo is developing in the uterus they increase to 15 mm, and may even grow right into the spiracle of the embryo. Interestingly enough, certain viviparous teleosts (bony fishes) have evolved almost exactly similar adaptations (see section 3.3, p. 22).

In the less advanced forms of viviparity the fetus does not differ greatly in structure and metabolism from that of a comparable oviparous species. Nutrition is by extracellular digestion of the contents of the yolk sac and the absorption of the products by the vitelline blood vessels. Gas exchange takes place at first over the entire surface of the body, then by means of the vitelline capillaries and finally by the gills. In some species gas exchange is facilitated by an increased gill area (Fig. 2–3). In the seven-

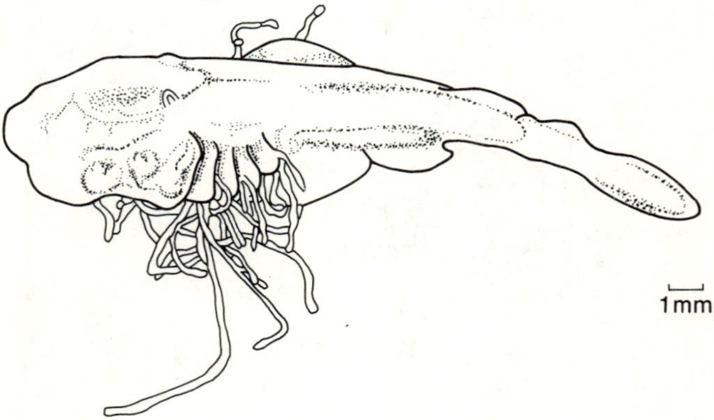

Fig. 2–3 Uterine embryo of *Torpedo*, showing greatly enlarged external gills. (From Grassé, P.-P. (1958), *Traité de Zoologie*, XIII. Masson, Paris.)

gilled shark (*Heptranchias perlo*), 184 separate gill filaments have been counted on each side of the embryo, while in *Urolophus* the external gills grow as long as the body. These adaptations indicate a low oxygen concentration in the uterus; which is also suggested by the fact that in some species the haemoglobin of the fetus has a greater affinity for oxygen than that of the mother. This will facilitate oxygen transfer from mother to fetus. It is an adaptation found also in mammals.

In some of the viviparous elasmobranchs, absorption of the yolk by the vitelline capillaries is progressively replaced by the passage of the yolk up a canal in the 'yolk stalk', which joins the fetus to its yolk sac, directly into the gut.

The vitelline capillaries, which absorb nutrients from the surface of the yolk mass, might also absorb nutrients from outside the embryo in the form of substances dissolved in the uterine fluid. This happens particularly in the viviparous rays (see section 2.3, p. 9). In many of the sharks and dogfish, the yolk sac lies in close contact with the uterine lining, and absorbs materials directly, as a true placenta.

The yolk sac placenta of the hammerhead shark (*Sphyrna tiburo*) is one of the few such types whose formation has been studied. About 3.5 months after fertilization the yolk sac makes contact with the wall of the uterus and the two surfaces grow to form interdigitating folds. Over the next few months these folds increase, as the yolk within the yolk sac is utilized. When the yolk is exhausted, the fetus is dependent on its mother. The layers of tissue between the maternal and fetal blood systems become greatly reduced in thickness, and the blood supply on both sides of the placenta is increased. Apart from the continued presence of the thin and membranous egg-case, the placenta resembles that of a mammal. Fig. 2–4 shows the yolk sac placenta of *Sphyrna* and a section of that of the smooth dogfish (*Mustelus canis*).

A further mode of providing nutrition for the embryo is found in the porbeagle (*Lamna nasus*). As in the placental species the yolk is completely exhausted early in development. Thereafter, the fetus swallows eggs—probably fertilized as well as unfertilized—as they reach the uterus. In this way a large amount of yolk passes directly into the gut of the fetus, whose stomach distends as a specialized 'yolk stomach', until its diameter may equal the length of the fetus itself (Fig. 2–5).

Primitive viviparity, with little or no nutritional dependence on the mother, does not require major adaptations by the fetus. In species where large amounts of nutrient are acquired, existing organs such as the yolk sac or the gut have become modified in function. However, extremely close metabolic relationships appear to have evolved with relatively slight modifications of the morphology of existing fetal or maternal organs. The extent of dependence on the mother of both placental and non-placental species is discussed in the next section.

2.3 Nutrition and growth

From the intimacy of the structural relations between mother and fetus, we might expect correspondingly close metabolic relations. On the basis of a considerable amount of work on a wide range of species, Ranzi divided the viviparous elasmobranchs into three groups according to the nature of the uterine secretions. These were described as predominantly serous, lipid or mucous.

Predominantly serous secretions are found in non-placental species, such as the electric ray (*Torpedo ocellata*) and the spiny dogfish (*Squalus acanthias*). In addition to substances diffusing directly from the blood serum, a proportion of the cells in the uterine epithelium secrete fat droplets into the lumen of the uterus. Some cells themselves migrate into the lumen, where they probably disintegrate and add to the 'uterine milk'. Typically, the amount of organic material in the fluid bathing the embryo is low; from 1 to 3%.

In some stingrays (*Dasyatis, Myliobatis*), a more complex secretion richer in lipids is found, with many cells present, and relatively large amounts of organic material; the uterine fluid of *Dasyatis violacea* has 13.3% organic

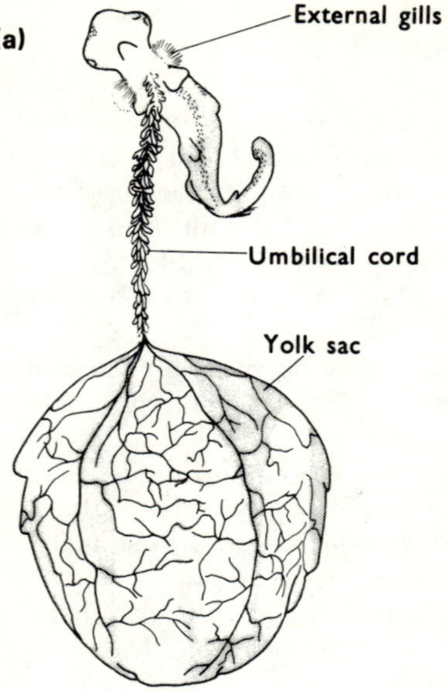

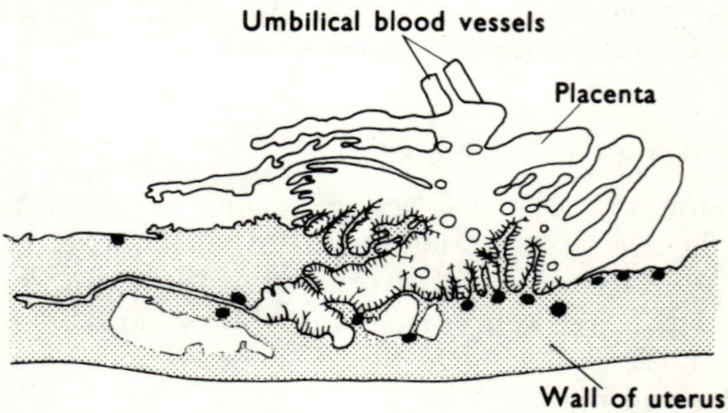

Fig. 2–4 **(a)** Yolk sac of *Sphyrna*; note large external gills and absorptive appendages on umbilical cord; **(b)** Section of placenta of *Mustelus canis* to show large surface area for absorption, and close relationship with uterine tissues. ((a) From Grassé, P.-P. (1958), *Traité de Zoologie*, XIII. Masson, Paris. (b) From Ranzi, S. (1934), *Pubbl. Staz. zool. Napol.*, **13**, 389.)

material, including 8.2% fat. The uterine wall in such species is thicker than in species with the other types of secretion, and more complex in structure.

The final type of secretion, mucous secretion, is found in a variety of dogfish and sharks, including species with and without placentae. Mucus is secreted by the uterine wall and taken up by the embryo, but the lipid content of the secretion is low and no cells are present. In the smooth dogfish (*Mustelus mustelus*), for instance, the uterine milk contains 5.1% organic material, but only 0.2% lipid.

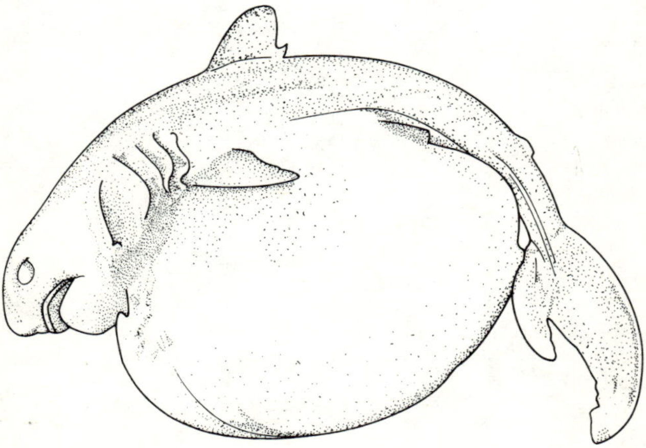

Fig. 2–5 Embryo of the porbeagle (*Lamna nasus*) with a greatly expanded 'yolk stomach'. (From Grassé, P.-P. (1958), *Traité de Zoologie*, XIII. Masson, Paris.)

A useful measure of the dependence of the embryos on these secretions is the increase in the amount of organic material in an embryo during gestation, since organic materials can have come only from the mother. In an oviparous species the amount of organic material declines by up to 20% during development, due to the utlization of stored nutrients and their removal as carbon dioxide and other waste products. Presumably much the same amount of energy is required for development of the embryos of viviparous species, hence a decrease of much less than 20% suggests some uptake of material from the mother, and an actual *increase* in organic weight indicates a considerable uptake.

Table 1 gives information concerning the uterine secretions of a number of species. Surprisingly, of the species with serous uterine secretions, only one (*Squalus blainvillei*) shows any net increase in organic material at all. In most cases, the amount of organic material taken up from the mother is insufficient even to offset the metabolic losses. In

Table 1 The differing compositions of uterine secretions in a number of elasmobranch species and the relationship between uterine milk composition, nutrient uptake by the fetus, and the effect on the maternal liver. The three types of secretion are serous (S), mucous (M) and lipid (L). In the case of *M. canis*, the maternal liver actually increases after declining by 37.4% (after Ranzi, quoted in NEEDHAM (1950).)

Species	*Type of secretion*	*Change in organic content of egg (%)*	*Organic material in secretion* (%)	*Decline in relative weight of maternal liver* (%)
Oviparous				
S. canicula		−21		
Viviparous				
non-placental				
T. ocellata	S	−23	1.2	8.5
T. marmorata	S	−31	1.5	9.1
Centrophorus granulosus	S	−54	1.8	
Squalus acanthias	S	−40	2.4	
S. blainvillei	S	+1	2.8	13.6
Mustelus mustelus	M	+369	5.1	56.1
Myliobatis bovina	L	+3120		
Dasyatis violacea	L	+1680	13.3	51.3
placental				
Mustelus canis	M	+1050	9.1	37.4

species with more than 3% organic materials in the uterine milk, the embryo does show uptake of material from the mother; in one species, the amount of organic material in the embryo increases over thirty-fold!

The principal metabolic store of the female is in the liver, and the extent to which this store is depleted should provide some measure of the impact of pregnancy on the metabolism of the mother. Table 2.1 indicates the percentage by which the weight of the maternal liver, in relation to body weight, declines during gestation. The greater the percentage of organic material in the uterine secretion, the greater the decline in relative liver size in the pregnant female.

Curiously, the presence of a placenta is not associated with the greatest increase in organic weight of the embryos. It may be that the evolution of a placenta is advantageous in more subtle ways. Unlike the situation in non-placental species, the liver of a female *Mustelus canis*, after an initial decline, actually *increases* in size during pregnancy. Perhaps, therefore, the placental form of viviparity is, for some reason, less metabolically demanding on the mother.

It is possible to read too much into Ranzi's data. For instance, the spiny shark *Centrophorus*, where the organic content of the embryo declines by the greatest amount (54%) during gestation, has an egg which initially weights 308 g (162 g organic material). Can we really compare this with an oviparous dogfish with an egg of 1.3 g (0.61 g initial organic weight) or with a stingray such as *Dasyatis* (initial weight 2 g, organic weight 0.9 g)? There may well be differences in metabolism or embryonic development in eggs of such vastly different sizes which make direct comparison invalid. However, in the absence of more precise studies on a range of species, comparisons of such crude parameters as percentage organic uptake is unfortunately the only approach possible.

2.4 Endocrine control

As in other vertebrate groups, the maturation of ova in elasmobranchs is controlled by the pituitary. Removal of the ventral lobe of this organ suppresses ovulation and causes developing eggs to degenerate; such regression has been recorded in several dogfish, both oviparous and viviparous. Some of the effects of hypophysectomy (pituitary removal) can be reversed by the administration of mammalian pituitary gonadotrophic hormones, which indicates their chemical similarity to the hormone which normally stimulates the elasmobranch ovary. Also present in the elasmobranch pituitary is a hormone with some of the properties of mammalian prolactin (see Chapter 6). No function has been established for this.

If, as in mammals, the elasmobranch pituitary controls the ovary, does the parallel go further; does the elasmobranch ovary in turn control the function of the uterus? An obvious experiment is to remove the ovary and see whether the uterus regresses or responds in any other way. This experiment has been performed but the only clear result was high mortality among the operated animals. An alternative approach is to inject an intact female with mammalian ovarian hormones, and look for an effect on the uterus. Injection of oestrogen (such as oestradiol-17β; see Fig. 2–6) into females of both oviparous and viviparous species is followed, after several months, by increased vascularization of the oviduct or uterus, and stimulation of the activity of the nidamental gland. This closely parallels the effects of oestrogen on a mammalian uterus.

Oestrogens have been detected in the ovaries and blood plasma of both oviparous and viviparous dogfish and rays, and the oestrogen level has been shown (in some species at least) to fluctuate with the reproductive cycle. It seems likely, therefore, that the uterus and associated glands are controlled by oestrogen secreted by the ovary.

Besides the developing ova and endocrine tissue, the elasmobranch ovary, like that of all vertebrates, contains a number of degenerating, or atretic, follicles. These arise from the remains of mature follicles after ovulation and from the large number of oocytes which start development but never complete it. In appearance, elasmobranch atretic follicles slightly resemble the corpora lutea (singular: corpus luteum) of mammals (see Chapter 6) and, like them, appear to be concerned with the metabolism of steroids. This suggests the hypothesis that, again like mammalian corpora lutea, the 'corpora lutea' of elasmobranchs secrete progesterone which maintains the uterus during pregnancy, possibly acting synergistically with oestrogen.

Progesterone is present in the ovaries of the viviparous *Torpedo marmorata* and *Squalus suckleyi* and, in the former species, varies in level

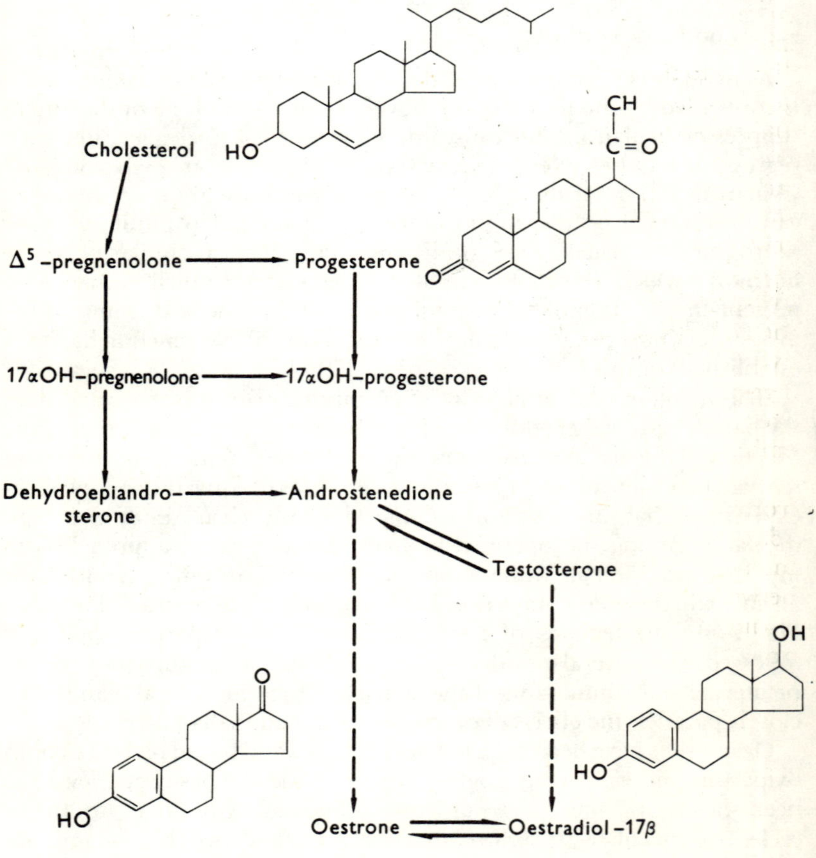

Fig. 2–6 The main pathways of steroid biosynthesis.

with the stage of gestation. In addition, the degree of lengthening of the folds of tissue in the uterine wall (see section 2.3) appears to correlate with the number of corpora lutea in the ovary.

However, many workers doubt that the 'corpora lutea' (atretic follicles) have any significance other than the removal of the remains of unwanted follicles and their contents. There is no *direct* evidence for any function of progesterone, and it may simply be a breakdown product of other, hormonally active, steroids, or an intermediate in their synthesis. A substrate that is a breakdown product or a metabolic intermediate in one species may of course evolve a biological function in another, and it would be unwise to generalize in a situation where the evidence is so sporadic.

A curious feature about some species, such as *Squalus suckleyi*, is that large quantities of oestrogen are present within the ova and in the embryos themselves, even at an advanced stage of development. This might act as a reservoir of the hormone which can be applied locally and directly to the lining of the uterus to maintain its high vascularity for the duration of pregnancy. Some such mechanism may be necessary, since in many cases the length of pregnancy much exceeds that of the normal non-pregnant reproductive cycle. It would be interesting to know whether large amounts of oestrogen are present also in the embryos of species where growth of the uterine lining is dependent on the presence of an embryo; for instance, *Urolophus halleri*, referred to in the previous section.

Differences between the endocrine control of reproduction in oviparous and viviparous species, therefore, appear slight. In both cases, the uterus or oviduct is under the control of oestrogen secreted by the ovary, and the ovary in turn under the control of the pituitary by means of a gonadotrophic hormone. Progesterone and 'corpora lutea', although present, are of dubious significance, and any effect of the presence of an embryo on the endocrine system of the mother is largely conjectural.

Despite the scanty evidence, it seems most likely that the transition from oviparity to viviparity occurred with no major endocrine adaptations, the most significant difference between the two being the interpolation into the reproductive cycle of a greatly increased time-lag between ovulation and deposition of the embryos. Until a great deal more is known of both oviparous and viviparous species, it is not possible to see how this might have arisen.

3 Bony Fishes

3.1 The forms of gestation

Viviparity in teleosts differs from that of any other group in that, almost without exception, fertilization occurs, and the young are gestated, actually within the ovary. In addition to producing eggs and hormones, the ovary has taken on the functions of a uterus.

Gestation within the ovary takes two forms. In one, the eggs are fertilized within the ovarian follicles. Development of the young proceeds within the follicle, and ovulation (if it can be said to occur at all) immediately precedes birth. This is known as 'follicular gestation'; it is typical of the family Poeciliidae, which includes such common aquarium species as the guppy (*Poecilia reticulata*), swordtail (*Xiphophorus helleri*) and platy (*X. maculatus*).

The alternative mode of gestation is 'ovarian gestation'; here the eggs are ovulated into the lumen of the hollow ovary (most teleosts have a single, hollow ovary situated centrally in the body cavity) and undergo development there. One example is the viviparous blenny (*Zoarces viviparus*) of British shores.

In addition, there are fish with intermediate forms of reproduction, for example displaying follicular gestation early in pregnancy and ovarian gestation subsequently; and also species with various more or less eccentric forms of gestation, such as the seahorse (*Hippocampus*) where it is the *male* which becomes pregnant, and a number of species where embryos are carried during development on the outside of the body (as in the greater pipefish, *Syngnathus acus*) or inside the mouth (for example, many common aquarium species of the cichlid family).

3.2 Development within the follicle

The structure of the poeciliid ovary is shown diagrammatically in Fig. 3–1. Sperm are deposited in the oviduct by the gonopodium of the male (a modified anal fin equivalent in function to the mammalian penis), and pass into the lumen of the ovary. Here they may either fertilize eggs immediately, or become embedded for storage in the cells of the epithelium which lines the lumen. Sperm remain viable within the ovary for many months and may derive nutrients from the female. Hence a female can continue to produce brood after brood in the absence of a male.

Development of the young within the ovary is not very different from the development of the embryos of related oviparous species. The only obvious structural modification is the great expansion of the pericardial sac (membranes which surround the heart) into two lateral flanges which grow over the head of the embryo and fuse together (fig. 3–1). The structure formed by the expanded pericardial membranes has a rich blood supply. For this reason, and because of its close contact (through the fertilization membrane) with the surrounding maternal tissues of the follicle, it is assumed to function as a site of exchange of materials between mother and fetus. It is referred to as a 'pseudoplacenta', mainly because of the feeling of some biologists that only the structure found in mammals is entitled to be termed a true placenta.

Presumably the mother supplies the developing embryo with the oxygen and water that it would otherwise obtain from the environment, and likewise removes carbon dioxide and nitrogenous wastes from the environment surrounding the embryo.

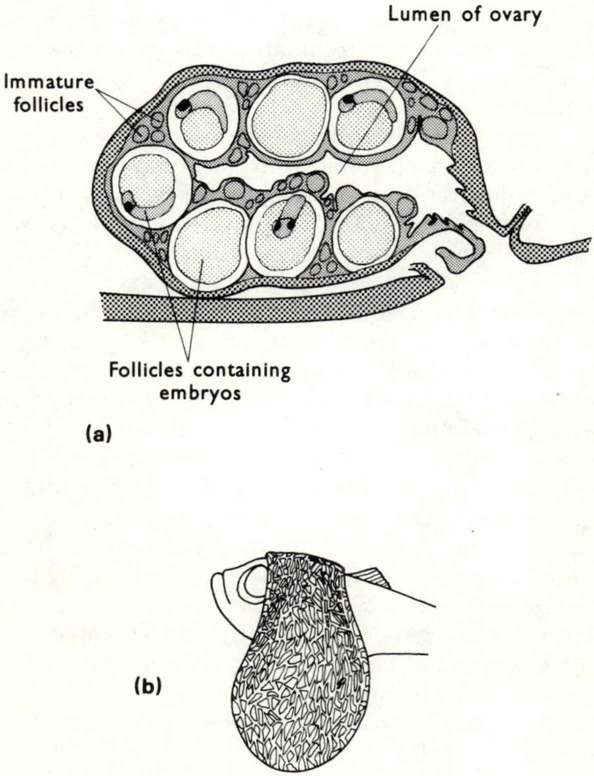

Fig. 3–1 (**a**) Schematic view of the ovary of a poeciliid, showing large follicles containing young. (**b**) An enlarged view of an embryo extracted from its follicle, showing the flanges of the pericardial sac which have met behind the head.

Direct evidence for the passage of organic materials into the embryo during gestation in the fish is sparse. If radioactively labelled substances of low molecular weight (such as glucose or amino acids) are injected into a pregnant female guppy or swordtail, they are later found only in the minutest quantities in the fetus. However, since much larger molecules, such as the high molecular weight iron-dextran complex 'Myofer', may pass into the *Xiphophorus* fetus quite freely, it would appear that the transfer of material is a selective process.

The importance of nutrient transfer from the mother is indicated as with the elasmobranchs, by measurement of the changes in organic weight of the embryos during gestation. Most poeciliids show very little change in organic weight, implying uptake from the mother just sufficient to offset the loss due to the embryo's own metabolism. A few species, however, show a striking increase. One such is *Heterandria formosa* (the mosquito fish), in which the dry weight increases approximately four hundred fold during gestation. Apart from the minute amount of yolk present in the *Heterandria* egg at fertilization, this species closely resembles *Xiphophorus* in structure. The poeciliid pseudoplacenta is clearly a most versatile organ.

During pregnancy, the liver of a female poeciliid undergoes changes very similar to those which result from starvation. However, although these changes in the liver coincide with pregnancy, they are probably not in fact due to the metabolic strain of supplying nutrients to the fetus, but rather to the rapid synthesis of yolk for the subsequent batch of maturing oocytes.

As far as the endocrine control of reproduction is concerned, the poeciliids conform to the general vertebrate plan of pituitary control of oocyte maturation, probably, as with the elasmobranchs, by means of a single gonadotrophic hormone. This has been established by hypophysectomy (removal of the pituitary gland) which blocks the development of oocytes and may cause their regression. Subsequent replacement therapy with mammalian gonadotrophins, of which luteinizing hormone (LH) is the most effective, restores normal development (see Chapter 6). The gonadotrophin-producing cells of the pituitary show a cycle of activity which corresponds to the cycle of maturation of successive batches of oocytes.

Like all vertebrate gonads, the poeciliid ovary is capable of producing steroids. Steroid-synthesizing enzymes are present, particularly in the granulosa cells which surround each developing follicle, and an ovary cultured *in vitro* can convert suitable precursor substances into the oestrogen, oestradiol-17ß

Demonstrating that steroids may be synthesized does not itself establish that they function as hormones. However, there is additional evidence to suggest a hormonal role for ovarian steroids in the poeciliids.

In the guppy, the epithelial cells lining the oviduct change their appearance in sequence with the successive cycles of oocyte production by the ovary. Removal of the ovary stops the oviduct cycle and causes regression of the oviduct epithelium; subsequent injections of oestrogen reverse some at least of the effects of ovariectomy.

Removal of the ovary also causes an increase in activity of the gonadotrophic cells of the pituitary, while the injection of oestrogen inhibits their activity. This is good evidence for a negative feedback relationship between the pituitary and endocrine tissues of the ovary (Fig. 3–2), with pituitary hormone stimulating oestrogen production, and oestrogen acting back on the pituitary to inhibit further gonadotrophin production. Oestrogen injection also induces the changes in the structure of the liver already described (p. 18).

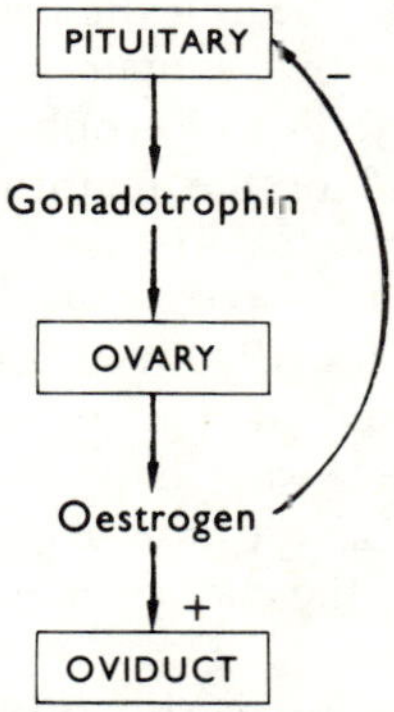

Fig. 3–2 Hormonal control of reproduction in a poeciliid. + = stimulation; − = inhibition.

The endocrine control of the ovary of viviparous teleosts in the production of mature oocytes and of oestrogen is very like that of oviparous teleosts. It is clearly necessary, therefore, to ask what additional rôle the hormones have during pregnancy.

Removal of the pituitary of a female early in gestation results in high embryonic mortality. This does not occur if the female is kept after hypophysectomy in a mixture of salt and fresh water, which indicates that the role of the pituitary is in maintaining a suitable salt/water balance of the fetal environment during gestation. The hormone responsible is prolactin: treatment with prolactin extracted from the pituitary glands of mammals is sufficient to maintain pregnancy in a hypophysectomized animal even in fresh water. The gonadotrophic hormone and oestrogen do not seem to play any part at all once pregnancy has started.

In an oviparous species, such as the ancestor of the poeciliids must have been, the pituitary gonadotrophin stimulates the maturation of the oocytes, as well as oestrogen production by the ovary. When the follicles are mature, they are stimulated to ovulate and the oocytes are released to the exterior and fertilized there. In some species, a rise in the level in the blood of both pituitary gonadotrophin and oestrogen has been shown to occur at the time of ovulation. As the injection of either mammalian luteinizing hormone or oestrogen results in ovulation in many oviparous species, it is reasonable to assume that the trigger to ovulation is a rise in one or both of these hormones.

The simplest model of how this system could have evolved to accommodate viviparity is that control of oocyte maturation remains identical to that in oviparous forms but, when the follicles mature, they do not immediately become sensitive to the ovulation-inducing effects of gonadotrophin or oestrogen. Instead, the oocytes are fertilized still within their follicles. Meanwhile, the next crop of follicles begins the final stages of their maturation. By the time the second batch of follicles is fully mature, the first batch—now containing advanced embryos—has belatedly become sensitive to gonadotrophin (or oestrogen). The rise in hormone levels at the end of the maturation cycle now causes the postponed 'ovulation' to occur. The young are released from their follicles (one cycle late) and pass to the outside world.

This hypothesis is easily tested; if true, LH and oestrogen injections into a female at an advanced stage of pregnancy should induce birth prematurely. This is, in fact, what happens. The stimulus to birth in viviparous species is identical with the ovulation stimulus in oviparous ones and the model, of how the endocrine system could have adapted to viviparity, stands.

The importance of the varying sensitivity of the follicle to the ovulatory effects of hormones is most obvious in *Heterandria formosa*; in this species, a female may contain, at the same time, six or seven broods of young, all at different stages of development. As the successive broods mature, each in turn is born. It is hard to see how this complicated situation could be controlled without each follicle controlling, to some extent at least, the timing of its own ovulation. If the timing of ovulation/birth was controlled by a single stimulus by the pituitary on follicles which were all equally sensitive, then presumably all follicles would empty simultaneously and most of the young would be born at varying stages of prematurity.

The ovary of the poeciliids has, like the mammalian uterus, an outer layer of smooth muscle, and contraction of this muscle layer probably moves the ovulated young out of the ovary at birth. The mammalian uterus contracts strongly if treated with the pituitary hormone oxytocin, or with a substance such as prostaglandin $F_{2\alpha}$ (Fig. 3–3); the same is true of the poeciliid ovary. This suggests that these substances, or similar ones, might be the natural inducers of parturition. In fact a substance very similar in structure to mammalian oxytocin (arginine vasotocin—see Fig. 3–3) is present in fish and has similar effects.

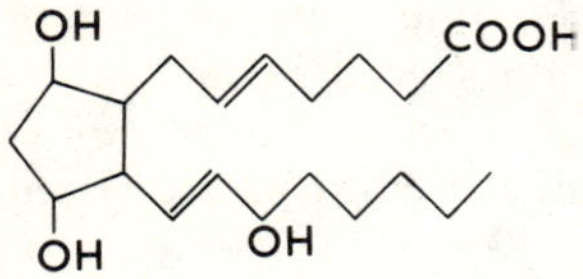

Prostaglandin $F_{2\alpha}$

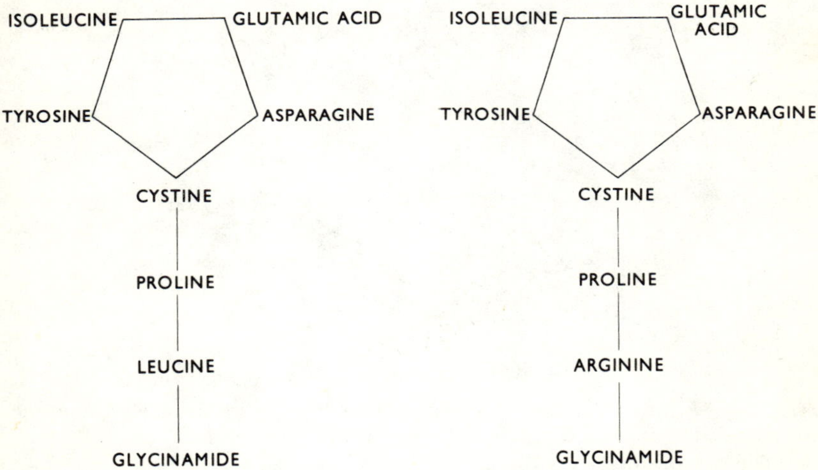

Oxytocin

Arginine vasotocin

Fig. 3–3 Structures of prostaglandin $F_{2\alpha}$, oxytocin and arginine vasotocin.

The question of how the embryo, considered as a foreign tissue graft, survives has been investigated only in the swordtail. Here the fertilization membrane appears to act as a barrier which may prevent access of fetal material to the mother and consequent sensitization. Alternatively, if the mother does become sensitized, the barrier must protect the fetus from immune rejection.

Follicular gestation occurs in a few other families of fish besides the Poeciliidae, mostly within the same Order, the Cyprinodontiformes. In the Anablepidae (called the 'four-eyed fish' because the eyes are divided into an upper half for looking out of the water and a lower half for looking under water) the pericardial sac also becomes highly expanded and functions as an organ of exchange with the mother. Here, however, it expands backwards, not over the head, to accommodate the very much enlarged small intestine (Fig. 3–4). The intestine appears to be fulfilling its usual function, the absorption of materials—but with the nutrients passing inwards from the exterior; the opposite direction from normal! About the other cyprinodont species with follicular gestation, very little is known.

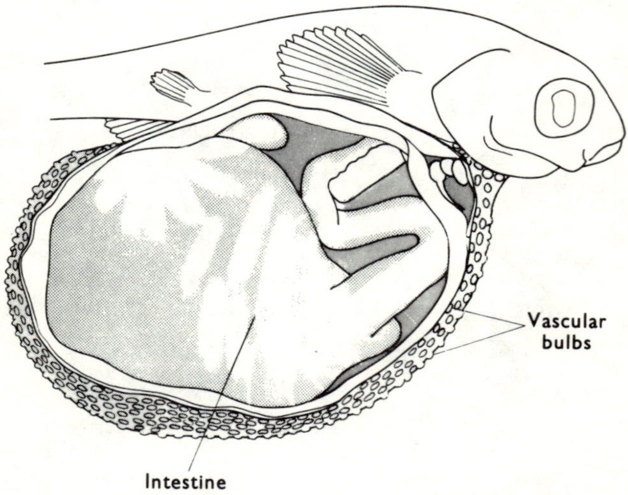

Fig. 3–4 Fetus of *Anableps* dissected to show the greatly expanded intestine surrounded by expanded pericardial sacs. (From Turner, C. L. (1940), *J. Morphol*, **67**. The Wistar Press.)

3·3 The ovary as a uterus

Two families of fish—related to the Poeciliidae and, like them, small freshwater fish from Latin America—show a mode of reproduction which indicates one way in which ovarian gestation could have evolved. In the Jenynsiidae, eggs are fertilized within the follicles. Initially, gestation is follicular and a pericardial pseudoplacenta forms. As the yolk supply is exhausted, the embryo loses its pseudoplacenta and is 'ovulated' into the lumen of the ovary, where it continues its development.

Although it has never been investigated directly, the rapid disappearance of yolk makes it seem likely that the embryo obtains nutrient direct from the ovary during this later phase of gestation. The epithelium lining the ovarian lumen grows out into a number of irregular flaps which penetrate the gill slits of the fetus and so enter its pharynx. Presumably nutrient is transferred by this means directly into the gut of the fetus; it forms a remarkable parallel with the flaps of uterine tissue which have a nutritive role in some of the viviparous rays (see p. 7 and Fig. 3–5).

The situation in the Goodeidae is similar, except that in this family it is the embryo, and not the ovary, which forms special absorptive organs. These trophotaenia—which again have close parallels in the elasmobranchs—are outgrowths of the hind part of the gut (Fig. 3–5). In some of the more primitive members of the family, they may serve only as organs of respiration, but in the more advanced species there is good evidence for a nutritional role as well.

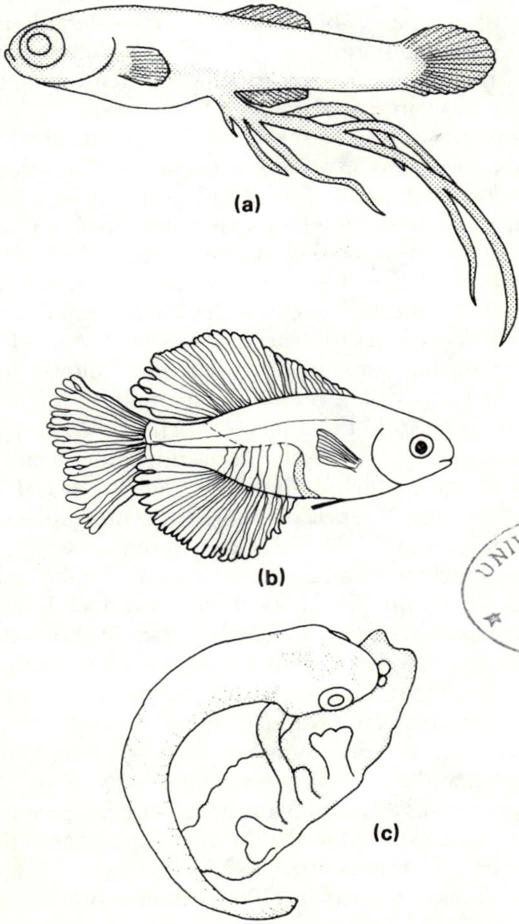

Fig. 3–5 Adaptations to viviparity in three teleosts. (**a**) Goodeid: trophotaeniae from hind gut. (**b**) Embiotocid: greatly expanded area of median fins. (**c**) *Jenynsia*: Flap of ovarian tissue grows in through gill aperture. ((a) and (c) from Turner, C. L., (1940), *J. Morphol.*, **67**. The Wistar Press. (b) from Webb, P. W. and Brett, J. R. (1972), *J. Fish. Res. Bd. Canada*, **29**, 1529.)

The ovarian gestation of other teleosts may have evolved by a different route, by the retention in the lumen of the ovary of eggs which had been ovulated and then fertilized there. At any rate, no transient follicular phase appears in gestation. Such forms include a heterogeneous collection of fish, belonging to several different families. Among these are coastal and deep sea marine species, some which live in fresh water, some with very small broods and some which produce thousands of young in each brood, some in which the fetuses are totally dependent on their yolk sacs for nourishment and some which get considerable nutrient from their mothers during gestation. As they differ so much in their physiology and ecology, there is no reason to suppose that any single universal advantage accrues from viviparity.

The very diversity of such forms serves yet again to stress the point that viviparity appears to have evolved several times within the teleosts, and the physiological adaptations found will probably be as diverse as the initial circumstances which made viviparity selectively advantageous.

Few species have been studied in any detail, and so any account can only deal with one or two selected examples, not necessarily representative in any way of other species with ovarian gestation.

More than fifty species of rockfish (*Sebastodes*) are found off the Californian coast, their habitats ranging from shallow pools on the sea shore down to depths of a few hundred metres. All are viviparous. Fertilization occurs within the lumen of each of the two paired ovaries. The young develop there for about one month before being released, still within their fertilization membranes. They then hatch and spend a larval period in the plankton. No specialized organs of nutrition develop during gestation, and relations between mother and embryo are probably restricted to gas exchange and the removal of metabolic wastes.

Fish with a planktonic larval stage often produce large numbers of offspring, and *Sebastodes* are no exception. In some of the smaller species, the brood size may be a mere 10^4 or so. On the other hand, the ovaries of one large female were found to contain an estimated 2.3×10^6 eggs. For a viviparous species, this is, to say the least, unusual.

The presence of such large numbers of developing young implies a considerable demand for oxygen from the mother. After ovulation, the rich capillary networks which previously surrounded each follicle remain intact, and are now exposed to the fluids of the lumen surrounding the embryos. Presumably this arrangement facilitates gas exchange with the embryos. In addition, the ovary itself has a double blood supply. Blood is delivered to both ends of the ovary by two separate arteries which join in a loop. This forms an efficient system for ensuring an even distribution of oxygenated arterial blood along the entire length of the ovary. In other animals, the ovary is served by a single artery; the double arterial system is apparently unique to the rockfish.

In contrast to the rockfish, the young of the viviparous surfperch

(Embiotocidae), also found in the eastern Pacific, are heavily dependent on the mother for nutrients during gestation. The dry weight of the developing embryo increases approximately seventeen times. This increase occurs partly as a result of the absorption of material dissolved in the ovarian fluid, partly by the swallowing of epithelial cells and other debris.

To facilitate the exchange of materials, both ovary and embryo show special adaptations. The ovary walls greatly increase in area during pregnancy, and become highly vascular and secretory. In late gestation, folds of the wall may even penetrate the gill cavity of the fetus and come into direct contact with the gills. This is reminiscent of the situation in the (unrelated) Jenynsiidae (see p. 22). Simultaneously, the absorptive surface area of the embryo is increased by spatulate extensions of the webs of the median and caudal fins, each with a rich blood supply (Fig. 3–5).

Such a large surface area would make it very difficult for the young animal to maintain its osmotic balance after birth. The difficulty seems to be avoided by a shunting of blood away from the capillary beds of the fin web extensions at birth. The consequent reduced blood supply to the expanded surface reduces the problem of osmotic control.

Little is known of the endocrine adaptations to viviparity in the surfperch, and even less in the rockfish. In the latter, as gestation lasts only thirty days, the situation may be similar to that in the poeciliids, with young born one complete cycle after the maturation and fertilization of a batch of eggs. The immunological problem of pregnancy has not been investigated in either group. Rockfish retain their fertilization membranes until birth, and these may constitute an immunological barrier. In the surfperch, the fertilization membrane is absent during gestation, and this cannot therefore be the explanation of embryonic survival. In fact, in one species of surfperch, *Cymatogaster aggregata*, antibodies certainly can pass from mother to fetus, and presumably give the embryo temporary passive immunity against disease. A similar process occurs in mammals.

3.4 Hippocampus—the pregnant male

One of the most bizarre forms of viviparity is that of the seahorse *Hippocampus*, in which the *male* becomes pregnant. Eggs are laid by the female in the normal way; the male fertilizes them and then takes them into a special ventral brood pouch (Fig. 3–6). Development proceeds there, and the young—of which there may be as many as 200 in a brood—are born a few weeks later. Possibly this should not be classed as true viviparity, but notwithstanding the unusual way in which the situation arises, it has a surprising amount in common with the more conventional forms of viviparity.

Apart from the seahorses, some related fishes show a similar, though less advanced, brood pouch. It is thus possible to follow the steps by which the situation in *Hippocampus* might have evolved. Many of the seahorses and the related pipefishes (and many other unrelated species as well) carry their eggs, during development, on the outside of the body of either parent. In the pipefish *Nerophis* and *Syngnathus*, for example (both quite common on British shores) the eggs are carried by the male. In the first case, they are simply attached loosely, in a mass, to the belly of the male; in the latter, embedded individually in spongy tissue on the belly. In some pipefish, as well as seahorses, the skin of the brood pouch forms a placental structure with a rich blood supply and, in some cases, a nutrient secretion.

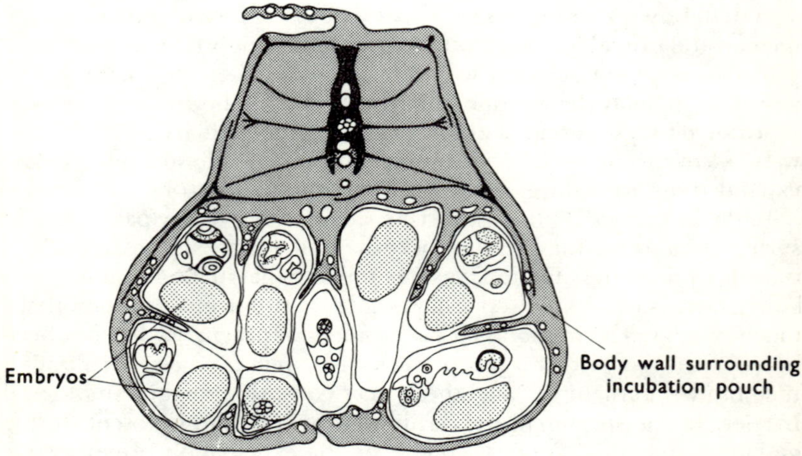

Fig. 3–6 Cross-section through a pregnant male *Hippocampus* showing a number of young in the incubation pouch. (From Grassé, P.-P. (1958), *Traité de Zoologie*, XIII. Masson, Paris.)

The development of the brood pouch is dependent on the presence of the male sex hormone testosterone, although if a sexually mature male is castrated his pouch does not degenerate for a long time. Gestation is not, therefore, immediately dependent on hormones from the gonad. What does appear to be necessary for gestation is the production, by the pituitary, of the hormone prolactin. Removal of the pituitary of a pregnant male seahorse causes high mortality of the young, and their premature delivery. As with the poeciliids, prolactin is necessary for the regulation of the ionic composition of the fetal environment during pregnancy. There is some evidence that the steroid progesterone is also required for gestation. If this is confirmed, the progesterone may be secreted somewhere other than the gonad, since castration has no effect on pregnancy.

Comparing this situation with more orthodox ones, it is interesting that, although it is the male and not the female that becomes pregnant, and although the target organs of the hormones are totally different, nevertheless two of the same hormones—prolactin and progesterone—are implicated in both cases.

3·5 Mouthbreeding

Some fish show a mode of reproduction which, while again it may not strictly speaking be classified as viviparity, is comparable to viviparity in many respects. This is the phenomenon of mouthbreeding, where the eggs are taken into the mouth or gill cavity of one or other parent after fertilization. Incubation takes place in the mouth and the young hatch there: in some species the fry may return to the parent's mouth later for protection.

Fish with this form of reproduction belong to four separate Orders and some are freshwater, some estuarine and some marine. As with the species showing true viviparity, the mouthbreeders are a heterogeneous group with little in common besides their method of reproduction.

In *Galeichthys*, an ariid catfish, embryos develop in the mouth of the male parent for about a month. The epithelium lining the mouth is modified in structure during this period; partly to hold the eggs in place, but it may also secrete nutrients or a protective bacteriocide. In some of the mouthbreeding cichlids a carbohydrate secretion has been identified histochemically, which may well be partly nutritive.

Galeichthys incubates a brood of up to 55 eggs, each 20 mm in diameter. In a mouthbreeding cardinal fish (*Apogon*), by contrast, the eggs are only 0.5 mm in diameter, but the male carries 20 000 at a time. Feeding by the incubating male must present problems: in some species there have been reports that the gut of the adult shrinks during incubation, while in the others eggs and embryos are occasionally found in the stomach or intestine. Although this has been attributed to 'intestinal incubation' (and even, presumably after inept dissection, to conventional viviparity!) it is more likely to be due to accidental swallowing of young by the parent. The only surprising aspect is how this is generally avoided.

4 Amphibia

4.1 Life on land

The Amphibia were the first vertebrates to spread on to the land. In most respects, for example, locomotion or sensory mechanisms, amphibians seem reasonably well adapted to a terrestrial life. In one major respect they retain a feature which imposes strict limits on the extent of their adaptive radiation on land. This is the presence of a moist skin, important for cutaneous respiration, but from which water evaporates freely. The resulting risk of desiccation means that, in general, amphibians are found only in moist places. Some have even reverted secondarily to a totally aquatic life.

A life on land presents problems also in reproduction. External fertilization is impossible, and the embryos and larvae need an aqueous environment in which to develop. Even the species which, as adults, are relatively independent of water, require to return from the land to breed.

Viviparity is, of course, a mode of reproduction which reduces this need to return to the water. If an adult amphibian can control its own water loss by physiological or behavioural means, then it can certainly maintain a suitable environment for an embryo developing within it. For this to be possible, internal fertilization is again a necessary prerequisite. Avoidance of desiccation of the young was probably, therefore, one of the principal advantages which resulted from the evolution of internal fertilization and of various forms of viviparity in the Amphibia.

The extant Amphibia are divided into three groups, all of which have some viviparous members. The Apoda (caecilians) lack limbs and generally live by burrowing underground. Not much is known of their reproduction; some lay eggs, but others are viviparous. One of the latter, *Typhlonectes*, is secondarily aquatic and lives in caves. The young have greatly expanded gills which serve as placentae, and appear to be nourished during gestation by uterine secretions.

The other two Orders of modern Amphibia are the Urodela (newts and salamanders) and the Anura (frogs and toads). Compared to the caecilians, both of these groups are successful and widespread. More is known of their reproduction, and accordingly they will be discussed separately in some detail.

4.2 Salamanders

Of the two divisions of the urodeles, the newts are exclusively oviparous, while the salamanders are either oviparous or viviparous.

Fertilization in both generally involves an elaborate courtship ritual in which the male deposits a spermatophore (spermatozoa enclosed in a protective capsule) which the female then picks up with the lips of her cloaca. Even in a species in which the male lacks an intromittent organ, internal fertilization is possible.

The salamander *Salamandra atra* lives in the Alps, at altitudes of about 800 to 3000 metres. Normally, it produces two live young, after a gestation period of about a year. At greater altitudes, where the annual period of hibernation is more protracted, gestation may last two or even three years.

Although the number of young born is only two, each ovary produces up to forty eggs; all of these are probably fertilized. Only the one lying furthest from the ovary in each oviduct (and probably, therefore the first to be ovulated) will actually develop; or, if this is removed experimentally, the next in line. The rest degenerate and in doing so form a nutritive solution for the embryos which do develop.

Young *Salamandra atra* closely resemble the young of oviparous salamander species, and pass through the sames stages of development. The only obvious morphological adaptation to viviparity is that the external gills of the larval stage are far larger than in comparable oviparous species. Probably this indicates an enhanced absorptive function, whether for gas exchange or nutrient uptake. At birth, the gills are resorbed and the young salamander takes on the adult form.

The oviduct (or uterus) appears to be under some form of endocrine control by the ovary. In both oviparous and viviparous species, removal of the ovaries results in reversion of the organ to a resting state, with loss of its rich blood supply and reduction in activity of its glandular cells. Corpora lutea are present in both, and in viviparous species their presence coincides with the time of gestation. There is evidence that they secrete progesterone. This does not seem to have any function in gestation, since ovariectomy of a pregnant female has no effect on her pregnancy. Possibly progesterone inhibits the development of a fresh crop of mature oocytes, since these start to increase in size only after parturition.

Viviparity seems to us, as mammals, a fundamentally different form of reproduction from oviparity; it comes as a surprise to find that in some species of salamander either live young or eggs may be deposited. The blind, colourless cave salamander *Proteus anguinis* (the olm) is an example. For many years, it was disputed whether this animal was oviparous or viviparous. Recent research indicates that it may be either. In laboratory conditions, female olms have been seen to produce both live young and eggs. Like *Salamandra atra*, many eggs (up to sixty) are ovulated but in normal conditons only one develops in each oviduct. If the temperature rises above 12°C or so, all of the eggs may be laid, although they will probably not all develop normally. Reproduction is therefore very flexible in this species, and possibly different reproductive processes can result from different environmental circumstances.

4·3 Frogs and toads

The only anuran to show true viviparity is the small toad *Nectophrynoïdes occidentalis*. This animal, which has no common English name, lives at altitudes of around 1200 metres in tropical West Africa, in an area with alternating wet and dry seasons. During the dry season, the toads avoid desiccation by remaining underground. Fertilization occurs at the end of one wet season, and birth in the middle of the next, which suggests that viviparity is an adaptation to avoid the exposure of eggs to dry conditions during development.

In the uterine regions of the oviduct, vascular villi are present (as in certain of the viviparous elasmobranchs and teleosts), and in the embryos the tail is enlarged, presumably as an absorptive organ. Otherwise, no major adaptations to viviparity are found. Despite the lack of a complex placental relationship, the developing embryos probably do obtain nutrient from the mother since at the time of fertilization the eggs are poor in yolk.

The endocrine control of pregnancy in *Nectophrynoïdes* has recently been worked out in some detail. The ovaries, like those of other vertebrates, contain structures resembling the corpora lutea of mammals, and which appear to be capable of synthesizing progesterone. Probably they are stimulated to do so by prolactin secreted by the pituitary. Removal of the ovaries from a pregnant female does not prevent gestation from proceeding, as would generally be the case in mammals. In fact the result of ovariectomy is that pregnancy proceeds more rapidly than in an intact animal and normal, fully developed young are born three or four months early. Conversely, injection of progesterone into castrated females slows down gestation.

This rather surprising relationship between ovary and uterus makes sense in the context of the annual reproductive cycle. During the dry season, when the adults are underground and inactive, the corpora lutea are active. This slows down embryonic development, and also suppresses the maturation of a fresh crop of oocytes. When the rain comes, the corpora lutea rapidly degenerate; simultaneously, larval development accelerates and the next batch of oocytes begins the final processes of maturation. Birth takes place and, a little later, ovulation of mature ova, mating, and a fresh conception.

Castration of a young female early in her first pregnancy is followed by abortion. This does not occur with the castration of multiparous females. As the corpora lutea of the older females are much more active in producing progesterone, it may be that in each pregnancy, there is a period early in gestation when progesterone is needed to prepare the

(a)

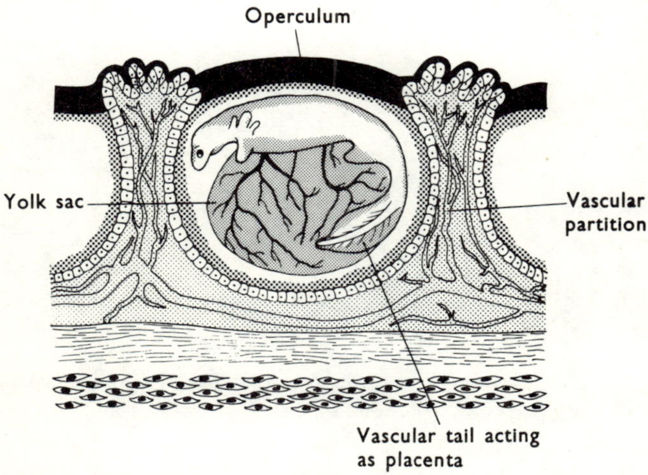

(b)

Fig. 4–1 (**a**) The Surinam toad (*Pipa*) carrying young. (**b**) An enlarged view of a tadpole developing in the skin of its mother's back. From Open University Course S321 *Physiology of Cells and Organisms*, Unit 9. Open University Press, Milton Keynes; and Parkes, A. S. (ed.) *Marshall's Physiology of Reproduction*. Churchill Livingstone, London.

uterus. In young females, with relatively inactive corpora lutea, a longer period of treament of the uterus is required; hence young females are likely to be more seriously affected by castration.

Like the teleosts, the anurans show some forms of external incubation besides the more conventional internal viviparity. In the Surinam toad (*Pipa surinamensis*), the fertilized eggs are embedded in cavities in spongy tissue on the back of the female. Each cavity is sealed by a lid. In this aqueous environment the young develop, absorbing oxygen and possibly nutrients from the highly vascular maternal tissues through its equally vascular tail (Fig. 4–1).

In another anuran, *Gastrotheca*, the eggs are carried in a specially formed pouch of skin on the back of the female. The larvae have highly expanded bell-shaped gills (Fig. 4–2) which are so large that they

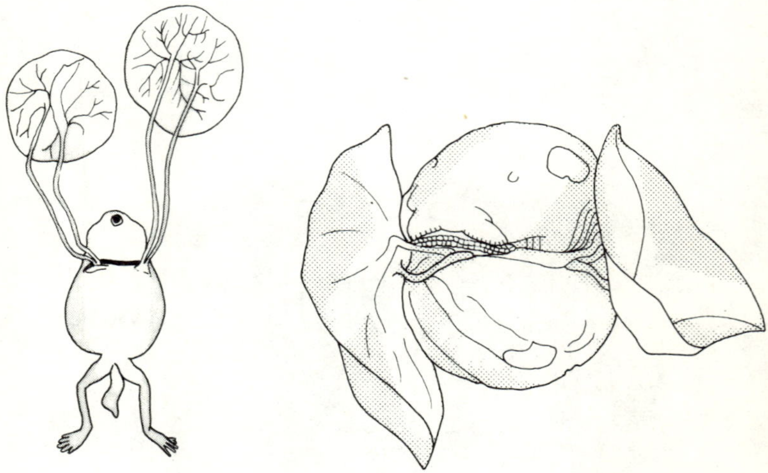

Fig. 4–2 Larvae of two species of *Gastrotheca*, to show the greatly expanded leaf-like gills. (From NEEDHAM, 1950).

completely surround the larva, and are closely applied to vascular ridges on the inside of the marsupial pouch. In one species of *Gastrotheca* the young are released from the pouch as tadpoles, in another they complete metamorphosis and emerge as frogs. Very little is known about the control of incubation in either *Pipa* or *Gastrotheca*, except that in the latter, oestrogen appears to control formation of the pouch. An Australian marsupial frog, *Assa darlingtoni*, has also been described.

Oral incubation also occurs in anurans; in the Chilean toad (*Rhinoderma darwini*) the male incubates the young in special mouth pouches, while one Australian anuran, *Rheobatrachus silus*, apparently uses a greatly expanded stomach for the same purpose.

In respect of their viviparity, brood pouches and oral incubation, the Amphibia show closely parallel reproductive adaptations to those developed in the teleosts. The evolutionary advantages for those parallel developments are clearly totally different in the two groups. In the teleosts, viviparity may confer various diverse benefits; in the partially terrestrial Amphibia, it seems reasonable to suppose that a single advantage, reduction of dependence on water for breeding, was obtained by the separate and independent evolution of several different forms of viviparity.

5 Reptiles

5.1 Problems of a waterproof egg

By virtue of the evolution of a skin which resisted water loss, and of an impermeable egg, reptiles were the first vertebrate group to be independent of the water for survival and reproduction. As in other fully terrestrial animals, internal fertilization, had to be developed, and reptiles all show some adaptation for this.

The enclosure of the egg in a virtually impermeable outer coating must be the last process of egg formation to occur before laying. Fertilization must have occurred some time previously, so even in an oviparous species, there will be a brief period after fertilization when the female contains developing embryos and not unfertilized eggs.

An embryo developing within a relatively waterproof (cleidoic) egg is faced with certain physiological problems. Gas exchange with the environment must still be able to occur. During embryonic development in reptiles, a longitudinal fold of tissue arises on either side of the embryonic axis. The two folds grow over the embryo and, by meeting and fusing in the mid-line, form two concentric membranes around it. The process is similar to the formation of the pseudoplacenta of poeciliids from two lateral folds of the pericardial membranes. In the reptiles, the inner of the two membranes formed is known as the amnion, and the outer as the chorion. Within the amnion is retained the amniotic fluid, which maintains the liquid environment of the embryo. The chorion becomes vascularized and closely lines the shell: it acts as an organ of gas exchange with the atmosphere.

As well as the gas exchange problem, the embryo within a cleidoic egg has a problem of waste disposal. Nitrogenous waste products cannot easily be eliminated. Aquatic animals generally produce soluble compounds such as ammonia or urea as the main end-products of protein breakdown, and as a means of getting rid of excess nitrogen. Both ammonia and urea are toxic and must not be allowed to accumulate; but they are also highly soluble and so easily disposed of by an aquatic animal.

In the enclosed environment of a cleidoic egg, an alternative mechanism is needed. Commonly, therefore, the principal end-product of nitrogen metabolism in reptiles is uric acid. This is virtually insoluble in water, and is therefore not toxic. The uric acid deposits are accommodated within a further accessory membrane, the allantois. The allantois lies close against the chorion, and because it is very vascular, it is also important in respiration.

The amnion, chorion and allantois, together with the yolk sac, are collectively known as the extra-embryonic membranes, because they function only as accessories during development and do not form part of the actual embryo (Fig. 5–1). As we shall see, they play a crucial role in the evolution of viviparity in the mammals as well as in the reptiles.

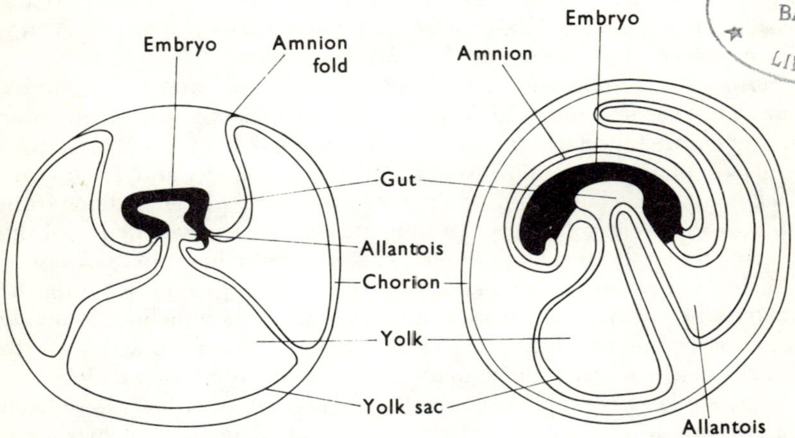

Fig. 5–1 Development of the extra-embryonic membranes in the egg of a reptile. (From Romer, A. S. (1970), *The Vertebrate Body*, 4th edn, W. B. Saunders, London.)

5.2 Relations with the mother

A primitive form of viviparity occurs in the viviparous lizard (*Lacerta vivipara*), common in Britain and Europe. The embryos develop within the oviduct much as they might outside it; in fact there have been reports that the viviparous lizard may occasionally be oviparous. No major physiological adaptations of the endocrine system have developed to permit a long time interval between fertilization and laying. The extraembryonic membranes must perform the same functions as in oviparous species, with the difference that gas exchange takes place with the uterine fluids and not directly with the atmosphere.

The yolk sac in oviparous species is an organ which digests and absorbs nutrients from the surface of the yolk mass within it, and transports the products of this digestion to the developing embryo. A viviparous embryo develops in the uterus, where it is surrounded by a medium containing nutrients secreted by the mother. A major physiological upheaval would probably not be required for the yolk sac to digest and absorb nutrients from the external medium rather than from its internal yolk mass.

A large number of viviparous reptiles have a yolk sac which does absorb material from the uterine fluids. In snakes such as the garter snake (*Thamnophis sirtalis*) and lizards like the European skink (*Chalcides tridactylus*), the yolk sac becomes closely applied to the lining of the uterus and forms a yolk sac placenta. The embryonic and maternal tissues form interdigitating folds which are separated only by the very thin residual egg membrane. The whole arrangement is closely similar to the elasmobranch yolk sac placenta (see p. 8); here again, two diverse groups have evolved similar structures in similar situations.

Other extraembryonic membranes, the chorion and the allantois, may also become involved in placental relationships. Chorioallantoic placentae are found in a variety of lizards and snakes, in conjunction with yolk sac placentae; again, two examples are *Chalcides* and *Thamnophis*. Details of the relationship of the chorioallantoic placenta differ from species to species; in many, an intimate contact is developed with the uterine lining. Both sides of the placenta may be reduced in thickness so that blood vessels of mother and fetus are in close apposition; in some the maternal capillaries are swollen, and the maternal epithelium is clearly secretory, while the chorionic cells may enlarge and in some cases form processes which penetrate between the cells of the uterine epithelium.

Few studies of the physiology of the placentae of reptiles have been carried out. In *Thamnophis*, newborn snakes have a higher protein content than eggs at the time of fertilization; about 45 mg protein must be transferred into each embryo in the course of gestation. Radioactively labelled glycine injected into a pregnant female appears later in the embryo, so amino acids can certainly cross one of the two placentae. The transfer is probably by the yolk sac placenta, which also seems to transfer sodium ions, and large molecules such as methylene blue (molecular weight=320). The chorioallantoic placenta may be more important in gas exchange, although diffusion of nutrients from the maternal blood is also possible.

Urea, and not uric acid, is produced as a nitrogenous waste product and transferred to the maternal blood for elimination. This is an adaptation to what amounts to a secondarily aquatic life during embryonic development.

Other viviparous reptiles may utilize the two forms of placentae differently; the information is not available to judge how typical *Thamnophis* may be. Viviparity is sporadically distributed within the reptiles. A range of forms of placentae is found within both lizards and snakes; even within a single genus some species may be oviparous and others viviparous. An example of this is the genus *Lacerta*; of the two British species of this genus, the sand lizard (*L. agilis*) is oviparous while the viviparous lizard (*L. vivipara*), produces live young. Viviparity probably evolved a number of times within the reptiles, and there is no good reason why the details of placental physiology should be closely similar in different groups.

What all viviparous reptiles do have in common is oviparous ancestry. Extraembryonic membranes, evolved in the context of the cleidoic egg, constituted a degree of preadaptation to viviparity. After the initial, relatively slight step of egg retention this preadaptation was exploited in various ways by different groups of reptiles.

5.3 Endocrine control

As the boundary between oviparity and viviparity is so indistinct, it would be surprising if the hormonal control of viviparous reproduction differed radically from that of oviparity. The physiological mechanisms involved are much the same in both cases: a cycle culminating in ovulation, preparation of the oviduct for the passage of eggs, appropriate sexual behaviour, and laying of eggs or birth of young.

In outline, one would expect the reproductive systems to operate on the same plan as in other vertebrate Classes, where the hormonal activity of the ovary as well as the maturation and ovulation of oocytes within it is under pituitary control, and the pituitary, via the hypothalamus region of the brain, is controlled by a feedback of steroid hormones produced by the ovary. Peripheral sexual organs, such as the oviduct, might be stimulated by gonadal steroids, and the cyclical activity of the whole system modulated by external influences, such as day-length or temperature, acting on the pituitary by way of the sensory system and the brain.

The sparse information which is available does fit in with this outline. Like fish and other lower vertebrates, and unlike mammals and birds, only one pituitary gonadotrophin has been clearly demonstrated (although recent research indicates that there may be two). The situation in the ovary is more complex. Many different steroids are present, as precursors, hormones and degradation products; and it is not possible to assign specific functions to specific substances. As usual, it has proved easier to detect the presence than to establish the function of most of the steroids. Oestrone and oestradiol–17β (see Fig. 2–6, p. 14) have both been found in reptile ovaries and both stimulate the oviducts of turtles and lizards. Progesterone has also been described in the corpora lutea of a number of reptile species, but its function is not clear.

Not many viviparous reptile species have been studied in detail. Events before ovulation, such as the production of mature eggs and the preparation of the genital tract, are controlled by similar mechanisms to those of oviparous reptiles. In the desert night lizard *Xantusia vigilis*, for example, oestrogen and progesterone are required to prepare the uterus for pregnancy.

Viviparous species might be expected to differ from oviparous ones in subsequent parts of the reproductive cycle. After ovulation, the empty ovarian follicles hypertrophy and form corpora lutea. At the same time, ovulation probably induces the inactivation of the enzyme systems which convert progesterone to oestrogen (see Fig. 2–6, p. 14); consequently the level of progesterone in the blood rises. This effect is more marked in viviparous than in oviparous species.

The risks of attributing functions to the luteal structures of lower vertebrates because of their physical resemblance to the corpora lutea of mammals have already been discussed (Chapter 3). However, reptiles differ from fish in that there is clear evidence for physiological effects of progesterone *before* ovulation in viviparous species. In viviparous reptiles, what might the function of the corpora lutea be during pregnancy?

There are two features of viviparity where progesterone might have a role: firstly, if gestation lasts longer than the normal, non-pregnant reproductive cycle, the ovulation of a fresh batch of eggs might be suppressed by progesterone; and secondly, the uterus must be maintained in a receptive state for longer where an embryo is retained during its development than for the rapid transit of an egg to the exterior.

In the viviparous blue spiny lizard (*Sceloporus cyanogenys*), ovulation and oviduct growth are suppressed by implanting oestrogen into the hypothalamus. This has not been demonstrated in an oviparous lizard, but might very well occur—many physiological mechanisms are controlled by negative feedback mechanisms of this sort. Perhaps of more specific relevance to viviparity, progesterone also inhibits ovulation. It does so by acting not on the hypothalamus or pituitary, but on the liver, probably by suppressing the oestrogen-stimulated synthesis of materials necessary for yolk secretion into the developing oocytes. The pituitary hormone prolactin also suppresses ovulation in *S. cyanogenys*, but how it does this is not clear. From our knowledge of its role in controlling the corpora lutea of some mammal species, an obvious theory is that it acts indirectly by stimulating progesterone secretion, but this does not seem to be the explanation.

Whether the corpora lutea are necessary to maintain a receptive uterus during pregnancy can be tested quite simply by removing the entire ovary some time after fertilization. In snakes, this sometimes results in abortion, and the ovary is probably essential at least for early pregnancy. In lizards pregnancy proceeds normally although the process of birth may be delayed or abnormal. Either the oviduct is independent of steroid support, or steroids come from some other source such as the adrenal cortex, or even from storage sites in the yolk of the embryos themselves. Immediately adjacent to the embryos in the uterus, certain histological changes do occur in the uterine epithelium during pregnancy, and it has been suggested that these are evidence of a local endocrine relationship between the embryo and the uterine mucosa.

Injecting progesterone to compensate for the loss of the corpora lutea did not reverse the effects of ovariectomy. In the lizard *Zootoca vivipara*, it delayed birth by a considerable period, a similar response to that seen in the amphibian *Nectophrynoïdes* (p. 30).

The principal relationships of the hormones to reproduction in viviparous lizards are summarized in Fig. 5–2.

As in fish and amphibia, contractions of the genital tract to expel the embryos at term are stimulated by arginine vasotocin (see Fig. 3–3, p. 21) secreted by the pituitary.

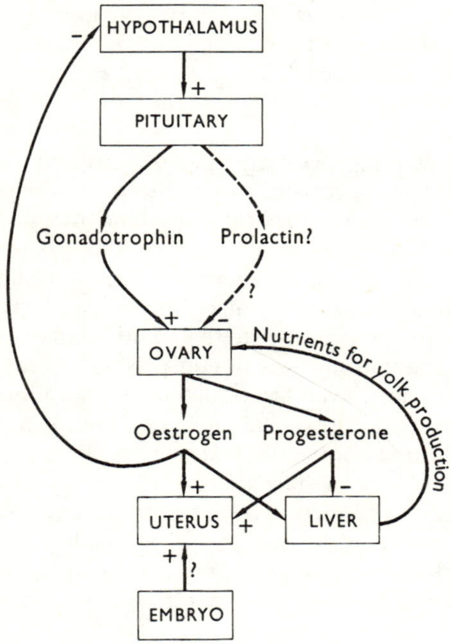

Fig. 5–2 Summary of the probable relationships between the reproductive organs in a viviparous lizard. + = stimulation; — = inhibition.

Only a very few reptiles, oviparous or viviparous, have been investigated, and it would be rash to generalize. However, it does appear that progesterone may be more important in reptiles than in lower groups, and that in viviparous reptiles it may be of significance in suppressing ovulation during pregnancy, and in influencing the time of birth. It is possible that embryos may have a direct and local action on the uterus. Apart from these rather uncertain differences, the transition to viviparity is not associated with any marked endocrine adaptations in the reptiles.

6 Mammals

6.1 Marsupials

The mammals are divided into three groups: the oviparous Monotremes (such as the duckbilled platypus), the Marsupials, and the Eutheria. These probably evolved separately, and do not represent successive stages of evolution on the way to the 'advanced' mammals such as ourselves.

Marsupials are now indigenous only in Australia and South America, although some have been successfully introduced elsewhere. In Australia they were until fairly recent times the only land mammals, and have diversified there into almost as wide a range of types as is shown by the Eutheria in the rest of the world. There are marsupial herbivores and carnivores, marsupial dogs, cats, mice, shrews, and even marsupial moles; all very similar in appearance to their Eutherian equivalents. With such a diversity of adaptations, it is not surprising to find a diversity in methods of reproduction. Even the pouch (or marsupium) which gives the group its name, is not universal; it is not found in the American opossums. Generalizing about marsupial reproduction is not easy, and it is made more difficult by the fact that in only a small proportion of marsupial species has reproduction been studied in any detail.

The structure of the reproductive organs of female marsupials differs from that of any other mammalian group. Although variations occur, a typical form is shown in Fig. 6–1. The most striking feature is the presence

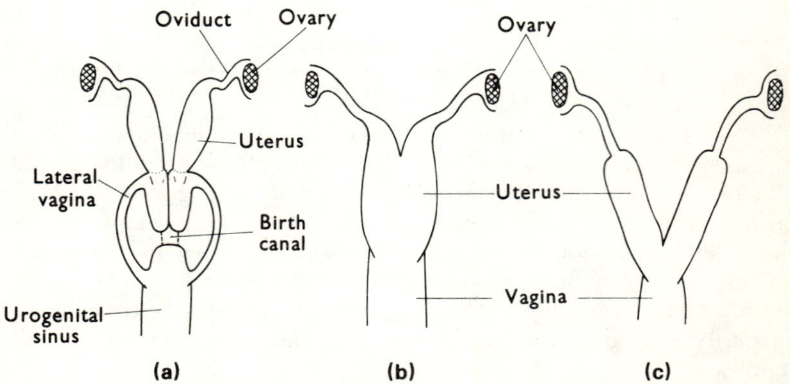

Fig. 6–1 Female reproductive system of a marsupial (**a**) compared with those of eutherian mammals (**b, c**).

of a double, Y-shaped vagina. Some male marsupials have a forked penis to match. The upper ends of each horn of the vagina enter a medial common chamber, which is also joined by the two horns of the uterus. Sperm ascend the lateral vaginal canals and fertilize eggs in the oviducts. The embryo implants in the uterine wall (endometrium) and develops there; when birth takes place, the blocked pseudovaginal canal opens; the fetus passes out, climbs into the mother's pouch, and continues its development there.

6.1.1 Relations between mother and fetus

The major placental organ in marsupials is the yolk sac. This expands greatly, almost entirely enclosing the developing embryo, and is therefore pressed closely against the uterine wall (Fig. 6–2). Part of the yolk sac is highly vascular.

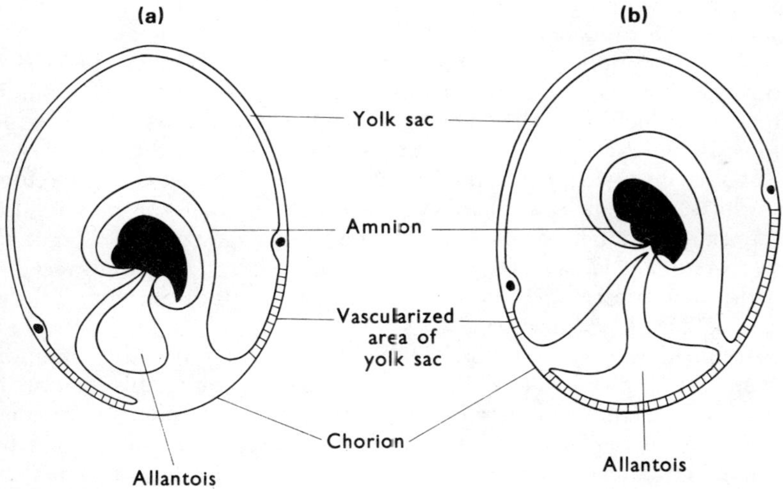

Fig. 6–2 Diagram to show the disposition of the fetal membranes in two marsupials **(a)** *Didelphis*, **(b)** *Perameles*. In *Didelphis* the allantois makes no contact with the chorion and has no placental function while in *Perameles* an elaborate chorioallantoic placenta forms. (After TYNDALE-BISCOE, 1973).

In some species, such as the North American opossum (*Didelphis marsupialis*), the fertilization membrane remains intact round the embryo throughout gestation; this limits the closeness of maternal–fetal contact. In other species, the membrane disintegrates and more intimate contacts develop. In the tammar wallaby (*Macropus eugenii*) and the four-eyed opossum (*Philander opossum*), microvilli on the yolk sac surface interdigitate with similar features on the uterine surface. At the boundary of the vascular and non-vascular regions of the *Philander* yolk sac, still closer contact is made, with fetal cells actually penetrating between uterine cells. As a result partly of this penetration, and partly of the thinning of the maternal and fetal epithelia, the two blood supplies are brought close together and exchange of material between mother and fetus is facilitated. In the native cat (*Dasyurus viverrinus*), fetal cells even penetrate capillaries of the mother and maternal blood enters the yolk sac itself.

In most marsupials the allantois is relatively small, lacks a rich blood supply, and is surrounded by the very much larger yolk sac. Most probably it functions solely to store excretory products. In a few species, however, it is much larger: in the long-nosed bandicoot (*Perameles nasuta*) the allantois grows into contact with the chorion and vascularizes it. In the area of contact between the two membranes, the chorion proliferates as a syncytium which erodes the uterine epithelium to expose uterine capillaries, forming a true chorioallantoic placenta (Fig. 6–2).

The relative importance of the different placental arrangements is hard to assess. It is likely that the vascular region of the yolk sac placenta is important chiefly in gas exchange, while nutrient uptake occurs across the non-vascular yolk sac membrane. In many species the allantois is probably involved mainly in the removal from the fetus of excretory products. The elaborate chorioallantoic placenta of *Dasyurus* is also important in nutrition. *Dasyurus* has the shortest gestation period of all the marsupials (12 days) and yet the young are relatively more advanced at birth than most marsupials: this may reflect the efficiency of the chorioallantoic arrangement.

At birth, the fetus is expelled from the uterus into the common chamber of the vagina. In a female carrying its first young, this chamber is a cul-de-sac; connective tissue forming the blocked pseudovaginal canal is actually torn as the fetus passes through. In most of the kangaroos, the passage remains open thereafter, but in almost all other marsupials it closes and is again forced open with each subsequent pregnancy.

The young animal crawls up the fur of its mother's belly into the pouch, attaches to a nipple, and continues development. The female comes into oestrus ('heat') at the time of birth, and will probably conceive again immediately, although development of the second embryo is suppressed as long as the first is still permanently in the pouch.

6.1.2 Endocrine control

Predictably, the maturation of the ova and their ovulation is under the hormonal control of the pituitary. No further action by the pituitary is needed to convert an evacuated follicle into a functional corpus luteum, or to maintain the corpus luteum once formed. Progesterone secretion by the corpus luteum maintains the uterus during pregnancy.

The lifespan of the corpus luteum is from twenty to forty days, according to species. This sets a limit to the length of the gestation period, since luteal support of the uterus is essential for pregnancy. Pregnancy can last only as long as the period of luteal activity of the oestrous cycle, and in most marsupial species is appreciably shorter than this.

The period of lactation is much longer than that of pregnancy. If the sequence of gestation, oestrus and fresh conception continued uninterrupted, several broods of young, all at different stages of development, would simultaneously be present in the pouch, and competing for the limited resources of the mother. This does not happen. In species where the young are born before the end of the cycle, the mother's return to oestrus is suppressed by the physical stimulus of her suckling young at the nipple. Only when the young animal is no longer attached to a nipple, and begins to leave the pouch periodically, is this inhibition lifted; pituitary gonadotrophin release occurs again, followed by ovulation, a return to oestrus and, probably, conception.

In the wallabies and kangaroos, the gestation period is as long as the oestrus cycle, and in one or two cases slightly longer. The young animal cannot reach the pouch in time to suppress further ovulation. Oestrus does in fact occur immediately after birth, and may well result in a fresh conception. In this case, when the conception of a new embryo cannot be prevented, its development is suppressed instead. The zygote divides until it consists of a hollow ball of about a hundred cells (blastocyst stage); it then enters a resting state known as 'embryonic diapause' which continues until shortly after the first embryo leaves the pouch for good.

In the red kangaroo (*Megaleia rufa*), for example, the first pregnancy lasts 33 days. The young kangaroo, or 'joey' is then attached permanently to a nipple for 235 days. The second conception occurs about two days after the end of the first pregnancy, and lasts for about 235 days; approximately 204 of which are spent in diapause and the remainder in development at the normal rate. The second birth therefore occurs just as the pouch becomes vacant, and will be followed by a third conception. The female kangaroo may therefore have three successive broods dependent on her at any one time; one young in diapause, a second permanently on the nipple, and the third and eldest returning periodically to suckle.

If the ovaries are removed from a female with diapausing young, diapause will continue. For the blastocyst to come out of diapause and resume development, either the ovaries or injections of progesterone (and sometimes oestrogen) are necessary. It seems likely that suckling induces the posterior part of the pituitary to release the hormone oxytocin (see Fig. 3–3, p. 21); apart from stimulating the release of milk from the mammary glands, this inhibits the corpus luteum and hence suppresses the development of the embryo.

The retardation of the return to oestrus may be controlled by a different mechanism. If the corpus luteum is removed from an ovary while a blastocyst is in diapause, diapause continues, but this time the female will return to oestrus. It is not known how this is effected, but since it suppresses oestrus behaviour, the corpus luteum cannot be as inactive during diapause as it appears.

In many respects the marsupials do not differ greatly from the reptiles in their mode of gestation. In both cases, yolk sac placentae are common, and chorioallantoic placentae less widespread. Endocrine control of oocyte maturation, ovulation, and support of the uterus are similar.

The major difference between the two groups is associated with the one characteristically mammalian feature of the marsupials which the reptiles totally lack: lactation and consequent nutritional dependence on the mother *after* birth. This has led to the evolution of mechanisms by which the postnatal development of one young is safeguarded by its ability to delay subsequent pregnancies. The developing marsupial can control the reproductive cycle of its mother, an ability which is found also in the Eutheria although largely by different means.

6.2 Eutheria

To a eutherian eye, the reproductive tract of the female eutherian mammal appears distinctly more straightforward than that of the marsupial (Fig. 6–1). Ova are released from the paired ovaries and enter the funnel-like proximal end of the oviducts. Fertilization occurs in the upper part of the oviduct, and the fertilized ovum usually begins development immediately. Around the time of entry into the uterus, the blastocyst stage has been reached. The embryo then implants into the endometrial layer of the uterus and continues development there until term.

The uterus may be single, as in humans, or bifid as in rats and mice. In either case, it communicates with a single median vagina. Unlike the situation in marsupials, the routes of sperm entry and of parturition are the same.

6.2.1 Relations between mother and fetus

Implantation of the embryo into the uterine lining is a highly complex process, and involves progressive interactions between the embryo and a uterus previously hormonally prepared to receive it. The nature of these interactions is now beginning to be made clear. Implantation in the Eutheria appears a more elaborate process than in the marsupials. The outer cells of the embryo (strictly speaking, of the conceptus: the cells which will form extraembryonic membranes as well as those which constitute the embryo proper) are known as trophoblast. These multiply vigorously and erode the uterine lining; as a response to the presence of the embryo, the adjacent uterine cells undergo a transformation known as the decidual cell response. From being loosely-packed fibroblast-like cells, they transform into rounded cells, closely linked together and containing relatively large amounts of glycogen. This transformation may play a part in nutrition of the early embryo, and may also be important in controlling the rapid growth of the potentially dangerous trophoblast. The result is, in either case, a physiological relationship between embryo and uterus generally closer than that found in the marsupials.

As in the marsupials and reptiles, both yolk sac and allantois may, with the chorion, form placental structures. In most cases, the chorioallantoic placenta is much the more important. Even with a chorioallantoic placenta, many differences in the gross morphology are found (Fig. 6–3),

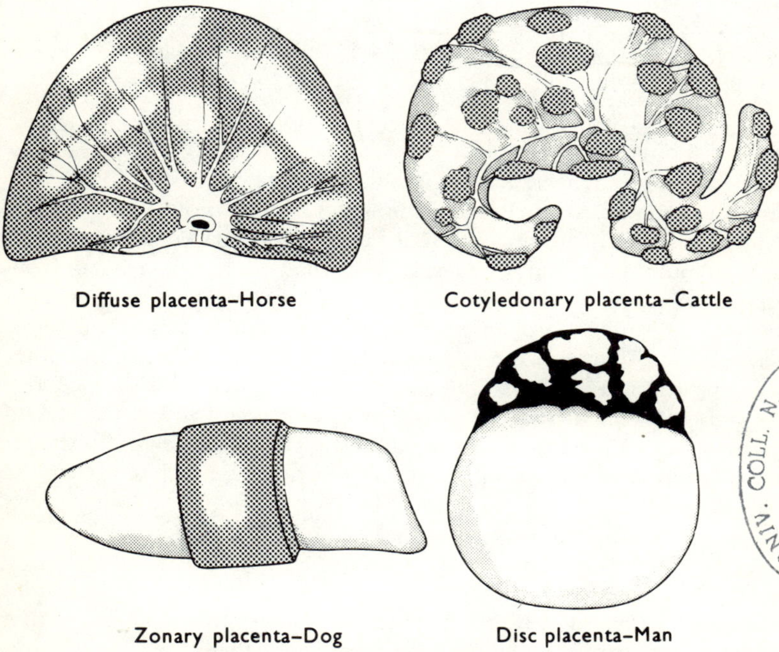

Fig. 6–3 Various forms of eutherian placenta. (From Witschi, E. (1956) *Development of Vertebrates*. W. B. Saunders, Philadelphia).

as well as in the degree of intimacy between maternal and fetal tissues. The maternal and fetal blood vessels are endothelial in nature, and are in primitive placentae each surrounded by connective tissue as well as by maternal epithelium or trophoblast. To pass from one blood circulation to the other, nutrients, waste products, gases and water would have to penetrate six layers of tissue: maternal blood vessel endothelium, maternal connective tissue, maternal epithelium, fetal trophoblast, fetal connective tissue and fetal endothelium.

In many species, one or more of these tissues may be missing and the physical contact between mother and fetus will be correspondingly closer (Table 2). It was thought that the fewer the intervening tissue layers, the

Table 2 Differences in the histological structure of mammalian placentae; + = present; — = absent.

	Maternal placenta			*Fetal placenta*		
Typical species	*endo-thelium*	*connective tissue*	*epi-thelium*	*tropho-blast*	*connective tissue*	*endo-thelium*
Pig, horse	+	+	+	+	+	+
Sheep, cow	+	+	—	+	+	+
Cat, dog	+	—	—	+	+	+
Man, monkey	—	—	—	+	+	+
Rabbit, mouse	—	—	—	—	±	+

more efficient the placenta; however, this is now considered a naïve view of the situation. In the first place, simple diffusion of molecules would be limited by the permeability of the least permeable barrier, rather than by the total number of barrier layers. Secondly, the placenta is far from being a passive physical filter. It is in fact highly selective in the transport of material. Amino acids, in general, are actively passed into the fetus, even against a considerable concentration gradient; urea, on the other hand, passes freely across the placenta by simple diffusion. Active transport requires metabolic energy, so that it is not surprising to find that the placenta itself uses a great deal of oxygen. On a weight for weight basis, the placenta may use up to two or three times as much oxygen as the fetus itself.

6.2.2 *Endocrine control*

A further, and major, difference between the eutherian mammals and other viviparous species (marsupials and lower vertebrates) is that pregnancy has an effect on the course of the reproductive cycle: ovulation and further oestrous cycles are suppressed for the duration of pregnancy. This does not occur in other viviparous vertebrates, where the gestation period is generally shorter than an unmated oestrous cycle, or at least shorter than the luteal phase of the cycle. Suppression of ovulation by pregnancy is, however, a feature we share with cockroaches (see p. 58).

In the marsupials, the mother does not respond to *pregnancy* with the suppression of the cycle, but the effects of *lactation* are similar. Prolongation of luteal life permits longer gestation periods in Eutheria than in marsupials, although it is not true that *all* Eutheria have longer gestation periods than their marsupial equivalents. The pregnancy of the shrew-like marsupial *Antechinus*, for example, is 25 to 31 days, longer than that of a comparable eutherian shrew.

Reproductive patterns in the Eutheria are highly diverse and varied, and the mechanisms of this maternal recognition of pregnancy vary considerably. A few examples, not representative of any species but their own, will illustrate this.

In the laboratory rat, the oestrous cycle is four or five days in length. Following ovulation, the corpora lutea form transiently, and do not become fully active. If a male is present, and fertilization occurs, the corpora lutea become fully active and persist for 20 or 21 days, the duration of pregnancy.

Whether the corpora lutea become fully functional or not cannot depend on the implantation of embryos in the uterus, since the zygote does not even enter the uterus until three days after ovulation, and it implants some time later. In an unmated female rat the corpora lutea have already started to degenerate by the third day from ovulation. The prolongation of luteal life must therefore be determined while the embryos are still in the oviducts.

This raises the question of the nature of the signal which indicates to the ovary that conception has probably occurred. In this case, the physical stimulus of copulation appears to be important. If an oestrous female rat is mated with a vasectomized (i.e. sterile) male, or even judiciously probed with a slender glass rod, she enters a state of 'pseudopregnancy'; the corpora lutea do not regress almost immediately as in the normal oestrous cycle, but function for up to twelve days, while the uterus undergoes many of the histological changes associated with the start of pregnancy.

Stimulation of touch receptors in the vagina or neck of the uterus causes the release of 'luteotrophic' hormones by the anterior pituitary. There appear, in the rat, to be two hormones involved; prolactin and luteinizing hormone (LH). The former is responsible, apart from its luteotrophic function, for preparing the mammary glands for lactation, while the latter—one of the two pituitary gonadotrophins found in the mammals—elicits ovulation and causes the initial formation of the corpora lutea from ovulated follicles in a process known as luteinization. As luteal maintenance is conditional on copulation, the corpora lutea become functional, and therefore the uterus receptive, before the embryos arrive ready to implant. Because this is contingent on copulation, and not on conception, there is the possibility of a 'wasted' pseudopregnancy; presumably this is rare in nature.

Corpora lutea maintained as a result of copulation last 12 days; the gestation period of the rat is 21 days. The difference between these two periods indicates a further extension of luteal life, this time conditional on pregnancy itself. If full luteal maintenance for 21 days is conditional on the presence of a conceptus (the fetus plus its extraembryonic) membranes), then it is reasonable to look for some stimulus emanating from the conceptus.

The rat placenta produces a luteotrophic hormone resembling prolactin. This supplements the pituitary luteotrophic hormones and permits the survival of the corpora lutea in the second half of pregnancy. In the last few days before birth, the placenta also produces progesterone. It is, therefore, a temporary endocrine organ as well as metabolic mediator between mother and fetus, taking over some of the functions of the pituitary and, later, the corpus leteum. The endocrine control of gestation in the rat is summarized in Fig. 6–4.

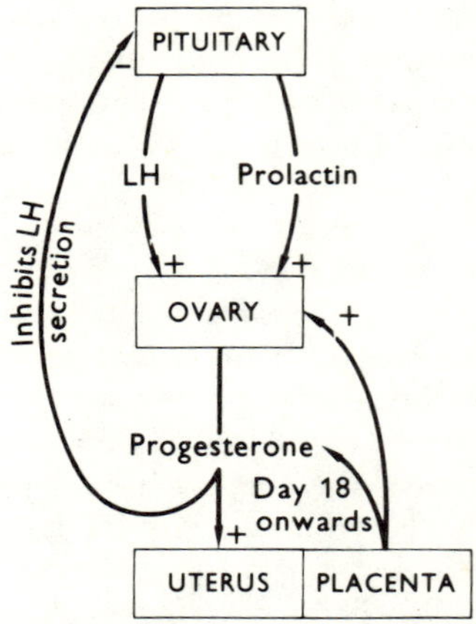

Fig. 6–4 Hormonal maintenance of pregnancy in the rat. + = stimulation; – = inhibition.

Extension of the lifespan of the corpora lutea of the oestrous cycle into the corpora lutea of pregnancy depends, in the rat, on two factors: copulation, and subsequent implantation leading to formation of a placenta. Copulation stimulates pituitary secretion of luteotrophic hormones, while the placenta produces placental luteotrophins and, for a while, leads to direct maintenance of the uterus by placental progesterone.

These two elements occur in many Eutheria, but with variations. In the ferret (*Mustela putorius*) ovulation as well as pseudopregnancy, is conditional on mating. In this species pseudopregnancy lasts as long as true pregnancy (42 days), so there is no need to postulate placental support of the corpora lutea.

The rabbit is another so-called 'induced ovulator' in which ovulation will not occur without copulation. Here the main luteotrophic hormones are prolactin and follicle stimulating hormone (FSH), rather than LH, together with oestrogen secreted by the ovary. FSH is the second of the mammalian pituitary gonadotrophins; as its name suggests it stimulates development and maturation of ovarian follicles.

Many Eutheria show a spontaneous period of activity of the corpora lutea during the oestrous cycle, whether or not copulation and conception occur. In sheep, for instance, the corpora lutea are active for 15 days in each unmated cycle. Here, too, the luteotrophic effects of prolactin and LH are necessary. This, however, is not the whole story. The sheep uterus also has a role in regulation of the normal oestrous cycle. If the uterus of a sheep is completely removed during the luteal phase of an oestrous cycle, the corpora lutea do not regress after 15 days, but may persist for up to 5 months. The gestation period of the sheep is 148 days.

The obvious inference is that the uterus produces some form of 'luteolytic' stimulus—one which positively terminates the life of the corpora lutea. Removal of the uterus removes the source of this stimulus; luteolysis does not occur at the normal time, and the corpora lutea can survive.

The luteolytic stimulus has been identified as prostaglandin $F_{2\alpha}$, one of a family of related substances with a variety of important functions in reproductive physiology. Prostaglandin $F_{2\alpha}$ is released by a progesterone-treated uterus, passes in blood vessels to the ovary and in some way, switches off progesterone production by the corpora lutea. If conception *does* occur, the presence of the embryo in the uterus must prevent prostaglandin release, or impede its action. The mechanism is not clear. However, it has been shown that infusing a macerate of sheep embryonic tissue into the uterus of a sheep prevents luteolysis while a similar infusion of pig embryo material does not; this suggests some form of species-specific chemical stimulus. Whatever the mechanism involved, the maternal recognition of pregnancy in the sheep appears to be primarily a matter of interfering with a luteolytic action, rather than of promoting a luteotrophic one.

The human menstrual cycle is similar to the oestrous cycle of the sheep in that a spontaneous luteal phase occurs, but prostaglandins do not seem to be important in bringing it to an end. Prostaglandin $F_{2\alpha}$ is certainly present in the uterus towards the end of the cycle. However, artificial treatment with prostaglandin earlier in the cycle does not result in premature regression of the corpus luteum and the onset of menstruation. Nor does hysterectomy (removal of the uterus) prolong luteal life. Hysterectomized women appear to continue normal cycles of ovulation, although obviously without a uterus they will no longer menstruate. The other difference between human cycles and those of most other mammals (other than primates) is that the human cycle is a menstrual one; the endometrium of the uterus undergoes a form of decidual cell reaction spontaneously at every cycle and the decidual cells are sloughed off at the end of the luteal phase of the cycle.

As with most other eutherian mammals studied, the human conceptus produces hormones which help to maintain the corpus luteum, the receptive condition of the uterus, and hence its own survival. One of the hormones is made by the trophoblast cells of the developing placenta from a few days after implantation. Because of its site of origin and function, this is known as human chorionic gonadotrophin (HCG). The advantage of using such unwieldy names is that the name describes the function. It is a protein hormone similar in chemical structure to the gonadotrophins produced by the pituitary and possibly having a common evolutionary origin with them. HCG is luteotrophic and probably prolongs the life of the corpus luteum to around 60 to 70 days.

In contrast to rats, implantation occurs in humans well before the normal demise of the corpus luteum, so a reprieve by the implanted conceptus is feasible. There is some evidence that HCG may also induce a second ovulation, and the formation of a second corpus luteum to take over as the first one declines. By whichever mechanism, prolongation of luteal function is achieved, conditional on conception and not, as in the rat, on copulation.

However, even with HCG production, the human corpus luteum of pregnancy regresses well before the end of pregnancy. The progesterone level in the blood does not decline as the corpus luteum degnerates. The placenta, which has already assumed the endocrine functions of the pituitary, now takes over the endocrine function of the corpus luteum and starts to produce progesterone, as well as oestrogen. The conceptus now controls its own gestation, and pregnancy will continue even if the ovaries and pituitary are removed. Not only are the maternal ovaries and pituitary now dispensable, but the placenta is autonomous in function so that even if the *fetus* were to be removed, the placenta alone would continue gestation and would be born at approximately the correct time. This is true of many species besides man.

As well as progesterone, the placenta also produces oestrogen during pregnancy. Together with progesterone, this is important in preparing the muscle layer of the uterus (the myometrium) for the powerful contractions which must occur to expel the fetus at term. Oestrogen promotes growth of the myometrium while progesterone suppresses its contractions. The roles of these two hormones in different species are probably similar, whether the steroids are made by the placenta, or whether the corpora lutea are active until the end of pregnancy.

The endocrine mechanisms controlling birth may differ in detail from species to species, but too few species have been studied to generalize about this. In all species which have been investigated the end of pregnancy is marked by a rise in oestrogen levels in the blood —sometimes an extremely steep rise—and a fall in progesterone. The fall in progesterone probably allows the uterus to contract if stimulated to do so, while high oestrogen levels facilitate rhythmic contraction.

At about the same time, prostaglandin $F_{2\alpha}$ rises sharply to a peak and, as it is a powerful stimulant of smooth muscle contractions, this is probably the main initiator of contractions of the myometrium. In the sheep, a positive feedback relationship between prostaglandin $F_{2\alpha}$ and oestrogen may develop. If this is the case (and the situation is not yet clear) it would account for the explosive rise in levels of both substances.

What initiates all these changes is not altogether clear, although in the sheep there is some evidence that the whole train of events is set in motion by the fetus itself. Whatever the details of the endocrine control of parturition, the result is the same; rhythmic contractions of the uterus, increasing in intensity until eventually birth occurs.

The uterus is not the only organ sensitive to placental or ovarian steroids during pregnancy. The mammary glands also develop during this period and mammary gland growth is dependent on oestrogen and progesterone, together with other hormones such as pituitary growth hormone (GH). Paralleling the situation in the uterus, oestrogen (working synergistically with GH) appears to prepare the glandular tissue to secrete milk, while progesterone prevents it from actually secreting. With birth, the progesterone level in the blood falls sharply, and the mammary glands can start active secretion.

Coinciding with the end of pregnancy, oxytocin secretion by the maternal pituitary accelerates. This may play a minor role in inducing parturition, but also stimulates the contractile tissue of the mammary glands to contract, squeezing the milk out of the milk ducts, and eventually to the nipple. The physical stimulus of the young animal suckling, as in the marsupials, elicits the reflex release of oxytocin by the mother, and hence expulsion of milk from the mammary glands.

In a normal, non-pregnant oestrous cycle, oestrous behaviour and ovulation are associated with a sharp rise in oestrogen production. Birth is associated with a similar oestrogen peak, so it is not surprising that many mammals, such as the mouse and rat, develop a 'post-partum oestrus'; that is, they come into oestrus immediately after giving birth. If a male is present, conception at the post-partum oestrus is highly likely.

This could result in the mother having to cope with the excessive demands of two simultaneous litters; one suckling and the other developing in the uterus. Kangaroos go to some lengths to avoid this situation, by imposing a period of diapause on the embryos in the uterus. A similar 'lactational delay of implantation' occurs in mice and rats. The length of the delay is roughly proportional to the number of young being suckled; in the rat, no delay at all occurs if the litter size is below five. Resumption of development can be induced at any time by an injection of oestrogen, so presumably the normal mechanism of lactational delay involves suppression of oestrogen production necessary for implantation. In other species (including man to some extent) ovulation is suppressed, and not implantation.

This account of the endocrine control of reproduction in eutherian mammals may appear complicated; it is in fact grossly oversimplified. In the human the placenta makes at least one further hormone which has a variety of effects on pregnancy and lactation. Also, of course, the few species discussed are typical only of themselves. The control of human gestation is not identical even to that of other primates, and the mouse and rat are not by any means representative of rodents as a group.

Any mode of reproduction has presumably evolved to maximize the number of viable young produced in the given circumstances. As the 4×10^3 or so species of eutherian mammals have, in other respects, evolved highly diverse adaptations to different circumstances, it is not surprising to find equal diversity in their reproduction.

Some, for example, show considerable delays in implantation, unrelated to the presence of suckling young. Badgers, seals and roe deer mate in the late summer. Birth of young in the middle of winter would be disadvantageous; the fertilized eggs reach the blastocyst stage, enter diapause, and so the gestation period is extended and birth takes place the following spring or early summer.

In the African elephant (*Loxodonta africana*), corpora lutea show only traces of progesterone and none at all is detectable in the blood during pregnancy. If gestation is not dependent on progesterone in the elephant, it is the only mammal where this is the case. Despite the apparent lack of progesterone, corpora lutea do seem to be necessary for pregnancy; not just a single one, but as many as forty may be present in each pregnancy. If a female elephant simply underwent up to 40 successive oestrous cycles, each of perhaps 3 weeks' duration, to accumulate enough corpora lutea to become pregnant, non-pregnant females with some corpora lutea present should be common. In fact, they are relatively rare. Possibly conception is preceded by a series of atypical oestrous cycles with multiple ovulation. The African elephant is not the most convenient of laboratory animals, so the question is not easy to resolve.

Just as eccentric is the plains viscacha (*Lagostomus maximus*), a South American rodent related to the Chinchilla. Here, from 200 to 400 eggs are ovulated at each cycle. About seven embryos implant, and of these all but two die and are resorbed before birth. The reason for this situation is unknown.

Only a small proportion of the mammalian species have been investigated in any detail. Presumably the diversity of reproductive patterns will seem even greater when more species have been studied.

Two features of eutherian reproduction stand out in the mass of complex detail. The first is that, unlike other viviparous vertebrates, means have evolved in the Eutheria of prolonging the lifespan of the corpus luteum, in the event of conception, beyond the length of the oestrous cycle. Long gestation periods are therefore possible and, compared with the marsupials, greater maturity of the young at birth.

Prolonged luteal life may be initiated by copulation, or occur spontaneously; the endocrine mechanism may involve various combinations of LH, FSH, prolactin and oestrogen, or the blocking of prostaglandin-induced luteolysis. Whatever the means, the end is the same. In other viviparous vertebrates, the corpus luteum determines the duration of pregnancy; in the eutheria pregnancy determines the duration of life of the corpus luteum.

The other obvious feature of eutherian reproduction is also a reflection of the declining importance of the corpus luteum. It is the tendency for control of gestation to pass progressively to the conceptus. The major role in controlling the corpus luteum is transferred firstly from the pituitary to the placenta; then the uterus is maintained directly by placental steroids. This relative autonomy of the conceptus, and in particular the crucial endocrine function of the placenta, is a feature peculiar to the Eutheria.

6.2.3 *The fetus as a graft*

One of the problems of viviparity referred to in Chapter 1 has scarcely been referred to since. This is the question of the survival of the fetus, considered as a tissue graft, in an immunologically hostile environment. The reason why it has not been discussed is our total ignorance of the situation in most fish, amphibia or reptiles. In mammals a great deal of effort has recently been put into elucidating the problem. If, despite this, no unequivocal answer emerges, at least it is because the situation is complex and not because it has not been investigated.

When foreign material—known as antigen—enters the body its destruction is mediated by specific antibodies. These are proteins (γ-globulins) which are either carried by white blood cells known as lymphocytes, or circulate in solution in the blood plasma. Immunization consists of an antigen inducing the formation of a type of antibody, or sensitized lymphocyte, which will react only with itself and not with any other antigen. The body is capable, therefore, of producing a staggering diversity of antibody types to deal with virtually any antigen to which it may be exposed. To simplify a complex situation, antigen injected in solution (for example a soluble protein) is destroyed by antibody also in solution, and a tissue graft by patrolling lymphocytes.

A fetus could avoid immunological attack by its mother by a variety of means. If it carried no antigens, no problem would arise. However, it has been shown that a mouse embryo carries antigens—including 'foreign' antigens inherited from its father—even before the time of implantation. So this explanation cannot be the correct one. Other explanations postulated are that the uterus is in some way a protected site where grafts of tissue can survive, or that the female cannot reject grafts effectively while she is pregnant. Both have been refuted as adequate explanations for fetal survival.

The fetus might be protected by a barrier which prevents sufficient contact with the mother for immune rejection to take place. There is some evidence in support of this idea. Fetal tissue transplanted to a site outside the uterus is destroyed; but not if it is transferred surrounded by an intact layer of trophoblast (see p. 44). In fact, an entire conceptus may thrive after transfer to many unlikely situations. In one experiment, it was shown that a mouse embryo would develop if inserted inside the testis of a male rat; the wrong site, the wrong species, and the wrong sex! Trophoblast could well constitute some form of barrier (in addition to all of its other function), since it is almost unique in apparently failing to express any antigens of its own.

Unfortunately for the hypothesis, there is now good evidence that the 'barrier' is permeable to antibody molecules, and even to maternal lymphocytes. In fact many fetuses obtain protection against diseases early in life by the transmission of antibodies from the mother before birth, so no complete barrier could exist.

If antibodies and lymphocytes can cross the placenta, why do they not destroy the fetus? It appears that this may be because the soluble antibody molecules attach to the antigens on the surface of cells of the fetus, masking them from recognition by the lymphocytes. Although lymphocytes may be present, they cannot recognize the fetus as being 'foreign' and cannot do any damange. The fetus survives by playing one half of the immune system off against the other.

Only a handful of species has been studied, and even in these the situation has not been clearly established. It seems reasonable to suppose that other species may have come up with other solutions to the immunological problems of pregnancy.

7 Invertebrates

7.1 Coelenterates

Viviparity is shown by one of the most familiar of the Coelenterata, the common beadlet anemone (*Actinia equina*). If disturbed, a beadlet anemone—whether male or female—may emit up to five or six juveniles from within its coelenteron. How do they come to be there? According to one theory, they are liberated into the plankton as tiny ciliated larvae which eventually enter an adult anemone (irrespective of its sex) and are incubated there until ready to emerge again as complete sea anemones.

Alternatively, it is thought that adults may occasionally change sex. While in a transitional state from one sex to the other, it is quite likely that self-fertilization will occur, since both male and female gonads will simultaneously be present. The young formed may then be incubated within the coelenteron of their own parent (which is, of course, both father and mother) without any intervening planktonic larval stage.

This would explain why, although adults differ in colour, extruded juveniles are almost always the same colour as the adult from which they came. If there is an obligatory planktonic stage in the life cycle, it is difficult to see how this would come about unless larvae choose to enter an adult of one colour rather than another (which is unlikely), or permanently adopt the colour of their foster-parent.

Although the young are present in close proximity to parental tissues, they are not attached in any way; no form of placenta is found. Judging from the fact that small fragments of crustacean limbs are often found within the guts of juveniles, they probably feed by swallowing portions of the prey swallowed once already by their parent. Presumably some mechanism operates to protect the young themselves from digestion.

7.2 Molluscs

Quite a few viviparous molluscs are known. Eight species of periwinkle (*Littorina*) are known from British shores, each species tending to occupy a different level on the shore. Of these, three species (*L. patula, L. rudis* and *L. neglecta*) are viviparous. All three are found in the upper half of the shore, which suggests that viviparity may be a means of avoiding the exposure of eggs to long periods of desiccation at low tide. On the other hand, the oviparous *L. neritoides* occurs so high on the shore that it is only reached by the sea at spring high tides. It is not known how this species solves the problem of releasing planktonic larvae into the sea with which it has such intermittent contact.

In the viviparous *Littorina* species, the young probably derive no nutrient from their parent. In other molluscs, this is not so. In a few cases (such as the prosobranch *Veloplacenta*) a placental relationship and nutrient transfer are suggested by structural adaptations of either embryo or parent, while in the freshwater river snail *Viviparus*, the young increase in size so greatly during gestation that uptake of organic material from the mother must occur.

7·3 Annelids

Among the annelids, viviparity occurs sporadically, and at least five families have viviparous representatives. In many, such as the syllid worm *Syllis vivipara*, gestation takes place in the coelom. In some cases, the nephridial canals become dilated and form uteri. The cirratulid *Ctenodrilus* has a form of placentation in which the early embryo is attached to the tissues of the mother and nourished directly through its blastopore. Subsequently it is released into the coelom and continues development there. In general, however, the developing young of viviparous annelid species are not nutritionally dependent on their mother.

It is not clear, in most of the viviparous annelids, what benefit is obtained from viviparity. In the case of one nereid worm, however, it is safe to infer a specific advantage. The oviparous *Nereis diversicolor* is restricted to the brackish water of estuaries, because although the adults can tolerate almost fresh water, the eggs and larvae die in fresh water and the species is dependent on the sea for breeding. The Californian *N. limnicola*, on the other hand, is internally self-fertilizing and viviparous. This eliminates the osmotically vulnerable early stages of development and greatly facilitates the permanent occupation of fresh water.

7·4 Onychophora and arthropods

In the arthropods and their near relatives, a startling range of forms of viviparity appears. The Onychophora, a phylum in some respects intermediate between annelids and arthropods, includes the celebrated *Peripatus*. This shows a complex form of placenta which has a superficial similarity to that of some vertebrates.

Among the Arthropoda proper, many crustaceans and arachnids produce live young. In the latter group, among the best examples are the scorpions, many of which are viviparous. In some, the eggs are rich in yolk and the young develop within the mother entirely at the expense of this yolk. In others, such as members of the genus *Scorpio*, there is a form of placenta, and the young develop within specialized uterine extensions of the female genital tract.

However, it is in the insects that viviparity occurs in the most remarkable range of forms. No fewer than thirteen orders of insects contain viviparous representatives, so viviparity is not just a rare and aberrant condition in an otherwise oviparous group. As in other groups of animals, the forms of viviparity vary widely; this makes comparisons difficult.

One group which has been much studied is the order Dictyoptera, the cockroaches. Many of these are oviparous, laying their eggs in batches, each batch enclosed in a tough protective capsule, the ootheca. Fertilization must, of course, occur before the ootheca is complete, so, as in vertebrates such as the reptiles, the eggs are fertilized and start developing before being laid. Even oviparous species of cockroach are therefore transiently viviparous.

In some species the ootheca is carried projecting from the female genital aperture for some time before being deposited while in others the eggs are carried by the mother until the young cockroaches hatch out. These species, such as *Blattella*, therefore show a rudimentary form of viviparity. A more advanced viviparity is shown in cockroaches such as *Diploptera* and *Leucophaea* where the ootheca is first extruded and then withdrawn again into a ventral brood pouch until hatching and birth.

During gestation, the embryos of many viviparous species increase in size, due in the main to water uptake from the mother. In one species, however, *Diploptera punctata*, the young increase several times in size during gestation; most of the increase is certainly due to water uptake, but as the dry weight also increases nutrients must pass from mother to embryo. No special structural adaptations are involved in nutrient transfer; the extensive pleuropodia (appendages of the first abdominal segment not found in adult insects) were at one time thought to have a placental function, but now seem more likely to be involved in splitting the egg membrane at hatching.

The sequence of different degrees of viviparity in the cockroaches parallels to some extent the array found in some vertebrate groups. Possibly similar selective pressures—such as the avoidance of desiccation of terrestrial eggs—operate in both cases.

Parallels exist also in the endocrine control of the processes of reproduction. In vertebrates, the reproductive organs are under control of hormones secreted by the pituitary; the pituitary in turn is controlled by the brain, by the neurosecretion of gonadotrophin releasing factors from the hypothalamus. In the insects, the gonads are largely under the control of gonadotrophic hormone secreted by the corpora allata (singular: corpus allatum). These structures, analogous in function to the vertebrate pituitary, are in an analogous position; close to the brain, and linked with it by the axons of neurosecretory cells (Fig. 7–1). The gonadotrophin in insects is almost certainly the same substance as that called juvenile hormone because of its effects in the control of sexual maturation, and its structure is shown in Fig. 7–2. There is some evidence for the existence of a second juvenile hormone with a similar structure.

The gonadotrophin has the effect of stimulating the growth of oocytes, in much the same way as the FSH of mammals (p. 48). In the cockroaches with more advanced forms of viviparity, mating itself stimulates egg maturation; in the most extreme case, that of *Diploptera*, maturation of eggs will not take place at all without copulation. This is presumably mediated by sensory stimuli transmitted to the brain, influencing gonadotrophin production by the corpora allata; the pathway, as well as the end result, has some similarity with the phenomenon of induced ovulation in the rabbit and ferret.

A female *Diploptera* may be pregnant for three months. During this time, no further eggs mature. As in the mammals, pregnancy inhibits the

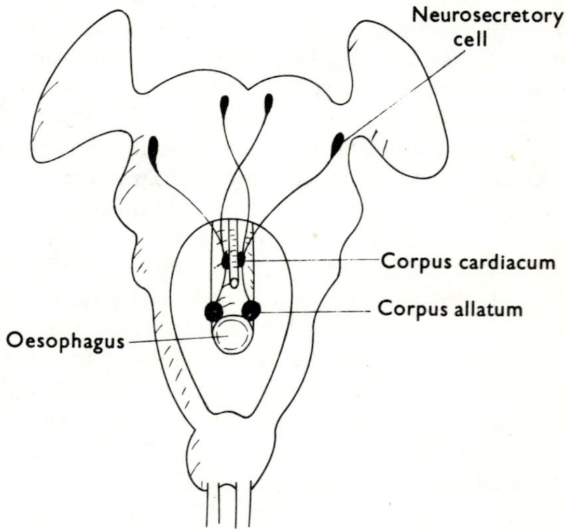

Fig. 7–1 Position of the corpora allata and neurosecretory cells on the brain of a cockroach. (From Chapman, R. F. (1969) *The Insects: Structure and Function.* Hodder & Stoughton Educational, London.)

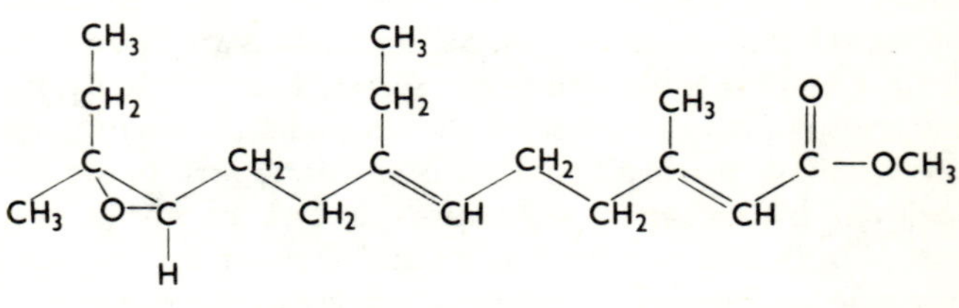

Fig. 7–2 Chemical structure of insect juvenile hormone.

production of mature eggs. Removal of the ootheca from the brood pouch of a pregnant female is soon followed by the maturation of a fresh crop of eggs; the continued presence of an ootheca is necessary for suppression of egg development.

This could be due either to some substance produced by the developing young, or to the physical distension of the reproductive tract by the ootheca. If a dummy ootheca is inserted in place of the real one, oocyte maturation is still suppressed, showing that the second explanation is probably correct. The effect of distension of the genitalia may be transmitted to the brain (and thence to the corpora allata) by neural, or hormonal mechanisms, or a combination of the two; the evidence at present is not decisive and differing mechanisms probably occur in different species.

Similar forms of viviparity are found in a variety of other species; for example the mayfly *Chloëon*, a few butterflies, some dipterans, and several beetles. Interestingly, several of the viviparous beetles are species which live inside termite or ant colonies. Possibly viviparity reduces the chance of the beetle egg being recognized as alien.

A different form of viviparity occurs in some diptera, such as the tsetse (*Glossina*) and some of its relatives. Here, the ovaries are very much reduced in size and each in turn produces one egg. This is retained in a distensible uterus and undergoes full embryonic development there.

After hatching—still in the uterus—the larva is nourished by the mother by a milk gland which secretes, through a nipple, into the uterus (Fig. 7–3). The relationship is reminiscent of that between the early marsupial young and its mother. Still applied to the nipple, the growing larva moults twice in succession. Finally, the third instar larva is deposited by the mother and pupates straight away.

Immediately after deposition of a larva, the next egg is ovulated, fertilized and starts development in the vacated uterus. This suggests that a similar mechanism may operate to that which suppresses egg maturation in the viviparous cockroaches.

The difficulty of moulting within the enclosed space of the uterus will be familiar to anyone who has tried undressing under the bedclothes with arms held by his sides. In the tsetse, this problem appears to be dealt with by a novel mechanism. Under the anterior part of the larva in the uterus lies a structure known as the choriothete. This is an adhesive pad with underlying muscle, and forms part of the wall of the uterus. It seems to attach to the egg membrane (which in insects is called the chorion; obviously a different structure from the chorion of the vertebrates). The chorion is then split by a tooth-like 'egg-burster' on the head of the larva, and stripped off by contraction of the choriothete muscles. This operation is repeated after each intrauterine moult to remove the successive larval moultskins.

Still more bizarre are the details of viviparity in insects of the order

Strepsiptera, which are endoparasites of other insects, chiefly bees and wasps. An example is *Stylops*, which lives in, among other hosts, the solitary bee *Andrena*. The adult male *Stylops* is winged and free-living, but the adult female is larval in form and, like the larvae of either sex, lives within the haemocoel of a host. A mature female pushes her cephalothorax—the unseparated head and thorax—out through the thin cuticle between two adjacent segments of the host's abdomen, while her own abdomen remains within. A male flies to her and mates by injecting sperm through the cuticle of the cephalothorax region, directly into the haemocoel of the female. The female then ovulates eggs directly into her haemocoel and fertilization occurs there. The embryos develop for some time in the haemocoel of their mother (which is, in turn, in the haemocoel of her host), and eventually pass through pores into a brood sac which is in fact the space between the outside of the mother's cuticle and unshed moultskins of the last larval and the pupal moults. Eventually, the embryos become larvae, pass up the brood canal and are born by an orifice just behind the mother's mouth.

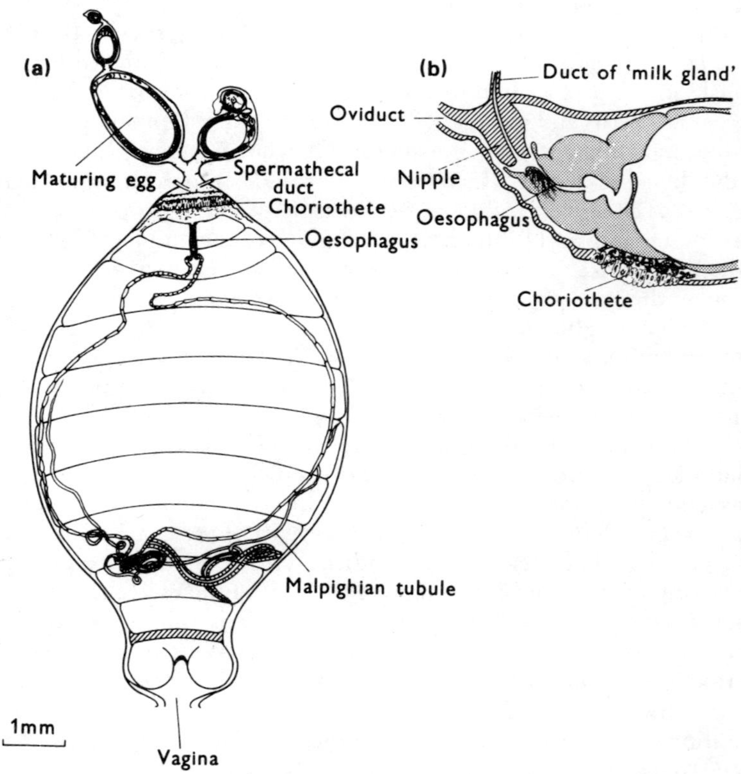

Fig. 7–3 Larva of tsetse *Glossina palpalis* in the uterus. (From ENGELMANN, 1970.)

As the female is still protruding from her host, the larvae emerge on to the surface of the bee. These first larvae have legs, so are able to leave the host and wait for a chance to transfer to a fresh one. This second bee then carries the larva back to its nest, where they enter the larvae of the bee and continue development, eventually either maturing in larval form as females or metamorphosing as adult males. The larval development of *Stylops* is synchronized with that of its host so that both reach maturity at the same time.

Very little is known of the physiology of strepsipteran viviparity. It may not be a coincidence that both embryos and larvae develop within the haemocoels of other insects; in the former case in the haemocoel of the mother and in the latter that of the host bee larva. Possibly adaptations with respect to nutrition are common to both stages of the life-cycle and an endoparasitic relationship with the bee host may have facilitated the evolution of viviparity, an endoparasitic relationship with the mother. However, this explanation is unlikely to hold for other insects, such as the gall midges (Diptera: Cecidomyidae), which show gestation within the haemocoel of the mother, but which are not endoparasites of other insects. In this case, larval as well as embryonic development takes place in the haemocoel.

Finally, a few insects show a truly placental arrangement during gestation. Of these, the best-known is *Hemimerus talpoides*, an aberrant earwig which lives as an ectoparasite in the fur of the giant rat (*Cricetomys gambianus*). The egg of *Hemimerus* is fertilized while still within its follicle, and develops there. Initially, the surrounding epithelium of the follicle is thin, but after fertilization it rapidly thickens, particularly at the two ends of the embryo. These thickened regions form the anterior and posterior maternal placentae (Fig. 7–4). The embryo grows in size and develops

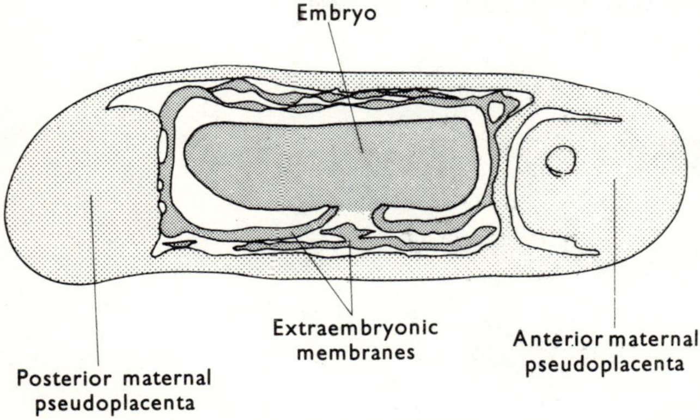

Fig. 7–4 Longitudinal section of the embryo of *Hemimerus* showing its placental relationship with the ovarian follicle. (From Hagan, H. R. (1951) *Embryology of the Vivaparous Insects*. Copyright 1951 The Ronald Press Company, New York.)

extraembryonic membranes which make close contact with the maternal follicular epithelium. In addition, certain cells of the embryo grow greatly in size and come to lie in close proximity to the anterior maternal placenta. Meanwhile, the maternal placental and other tissues of the follicle begin to break down. Presumably the resulting nutrients are taken up by the embryo, although this has not definitely been established.

Similar forms of placental viviparity occur in other groups of insects, such as the aphids (Aphidoidea), book lice (Psocoptera) and plant bugs (Heteroptera), although the relative importance of the different placental structures is not the same in each case.

Insect viviparity encompasses too great a diversity of types to be considered as a simple, uniform phenomenon. In some forms—for instance, the cockroaches—it may be comparable with vertebrate viviparity; the adaptive significance of viviparity in both cases may be related to the problem of avoiding desiccation, and some analogous physiological mechanisms appear to have evolved.

In other respects, insect viviparity is an adaptation to totally different circumstances. Many of the viviparous insects are parasitic; the Strepsiptera, *Glossina* and *Hemimerus*, have already been mentioned, but there are many others such as the hippoboscid flies (for example, the sheep ked *Melophagus*) and the hemipteran *Hesperoctenes* which is an ectoparasite of bats. Parasites often require specialized reproductive mechanisms to ensure dispersal of their offspring to fresh hosts. It may be advantageous, in the circumstances of a parasitic life, to give birth to a mobile larva which can find a host and become parasitic in its own right, rather than to lay an immobile egg.

Whatever the value of viviparity in any particular case, it is safe to assume that the diverse forms of viviparity in insects are strategies to cope with an equally diverse range of circumstances.

8 Conclusions

8.1 Diversity of viviparity

From its taxonomic distribution, as well as the diversity of forms in which it occurs, it is evident that viviparity must have evolved many times within the animal kingdom. In insects each Order with viviparous members represents a separate occurrence of viviparity; in some Orders, such as the Diptera, the existence of two very different forms of viviparity indicates that viviparity arose more then once within the Order. Similarly, in the teleosts, viviparity evolved several times; in the elasmobranchs, the sharks and rays show two lines of evolution of viviparity which may reflect separate origins; the viviparity found in the marsupial frogs bears no resemblance to that of other amphibia or any other animal. Amniote viviparity (reptiles and mammals) does show signs of a consistent pattern, but there is no reason to believe that this indicates a common viviparous ancestor.

The number of times that viviparity has arisen suggests that the transition from oviparity, when it occurred, was not a major one; that the species which evolved viviparity were to a large extent preadapted for it. One obvious preadaptation is internal fertilization, without which internal gestation is almost impossible. Internal fertilization is fairly widespread among animals which subsequently lay eggs, as well as among those which retain their young until birth.

Once internal fertilization was established, the fertilized eggs could be retained for progressively longer periods of time: even until the young were capable of looking after themselves. Initially, this retention of the eggs might require little or no physiological adjustment on the mother's part, beyond the separation in time of ovulation and deposition of the eggs. In some cases (e.g. cockroaches and eutherian mammals) the development of viviparity culminated in an extended gestation period made possible by suppression of ovulation during pregnancy. Maternal recognition of pregnancy, usually mediated by endocrine mechanisms, is a milestone in the evolution of more sophisticated forms of viviparity.

At an early stage in the evolution of viviparity, a retained egg would contain, at fertilization, sufficient yolk to complete embryonic development. If the embryo is to be retained, however, nutrients can be supplied after fertilization as well as before. There might be physiological benefits to the mother in supplying the nutrients over an extended period, although the absolute amount of nutrient will be much the same in either case. In some cases (poeciliid fish, for example, and the earwig *Hemimerus*) the same maternal tissue—the epithelium of the egg follicle—which supplied nutrients to the oocyte before fertilization later supplies nutrient to the young animal during gestation. Generally, the more nutrients supplied after fertilization to the embryo, the less yolk is present in the unfertilized egg; and the more complex the placental relationship.

The poeciliid teleosts and their relatives show a range of types which can be placed in a series which probably reflects the course of evolution of poeciliid viviparity. An oviparous species such as *Oryzias* (the medaka) shows a primitive form of internal fertilization, and lays its eggs almost immediately after copulation. Rarely in *Oryzias*, fertilization may occur while the eggs are still within their follicles; at least one pregnant *Oryzias* has been reported. Although this is obviously a pathological case, the embryos were still capable of development within the ovary, as in the poeciliids.

Poeciliids like the swordtail (*Xiphophorus helleri*) normally show fertilization of eggs within their follicles (although, abnormally, undeveloped eggs may be laid). Very little, if any, nutrient passes into the eggs during pregnancy and the eggs are large and contain much yolk. In *Heterandria*, another poeciliid, the embryos obtain large amounts of nutrient from their mother; at fertilization the eggs are small and contain little yolk. A similar sequence could be given of types of viviparity in the cockroaches and in other groups.

Comparisons of this sort are possible only within fairly homogeneous groups. Fossil evidence is of virtually no help in studying the evolution of reproductive mechanisms. As far as viviparity is concerned, the only significant contribution of palaeontology is the observation that adult Ichthyosaur skeletons sometimes contain the skeletons of very small Ichthyosaurs. These reptiles were therefore viviparous—or perhaps habitual cannibals!

Apart from this, in trying to reconstruct the course of evolution of viviparity, we are restricted to our knowledge of existing species. Information on structure is relatively easy to obtain but impossible to analyse without a knowledge of function. Our knowledge of the physiology of most viviparous animals is sparse. Until we have a great deal more information it will not be possible to do much more than speculate on how the mother and embryo in viviparous species learned to live with each other.

8.2 Viviparity as adaptation

Why might the evolution of viviparity have proved beneficial? Before viviparity could evolve, internal fertilization must first have been present. What advantages does internal fertilization confer? Clearly, it is important for land animals, since spermatozoa can survive only in an aqueous medium. But why should many species of fish show internal fertilization?

A complex courtship, leading up to mating, serves as a barrier to hybridization, because only a male of the correct species will make the appropriate behavioural responses and achieve fertilization. Intromission and internal fertilization are more difficult to achieve successfully, and may thus represent a further bar to mating between species. Another possibility (for which there is some experimental evidence) is that spermatozoa will not survive in the genital tract of a female of the wrong species, so again, internal fertilization will reduce the likelihood of interspecific mating being successful. Simple economy of sperm is unlikely to provide an explanation, since species with internal fertilization still produce a vast excess of sperm.

For whatever reason, internal fertilization has evolved in numerous cases. Some of the species with internal fertilization then interposed progressively longer delays between fertilization and deposition of the fertilized eggs. What benefits might accrue from this? One or two of the more obvious have already been discussed; for example the specific case of the urea dependence of elasmobranchs and the problem of the avoidance of desiccation in the embryos of terrestrial animals. In other cases, other factors may be important; for example, retaining the eggs may enable the female to be more selective in choosing where they are eventually deposited.

Perhaps viviparity was initially an extended form of the parental care of eggs and young found in, for example, a number of oviparous fish. In an environment where the oxygen content of the water is low, many fish aerate their eggs by fanning water currents over them. The same result can be achieved by carrying the young in the body of the mother, so that the embryo overcomes the limitations of its own respiration system and lack of mobility by employing the respiratory system and swimming ability of its mother.

For a young animal during gestation, the mother can be regarded as a buffer against the vicissitudes of the environment. This will be especially valuable in circumstances where the environment is a fluctuating or an unpredictable one. If the temperature, for example, varies widely, the mother will be able to reduce the range of variation of her own body temperature—and that of her gestating young—by behavioural or physiological means. Similarly, if food is in short supply, an adult will have a better chance of survival than a much smaller, free-living larva with no significant food reserves. A pregnant female will be able to supply nutrients continuously to her gestating young, while obtaining them only intermittently herself, and in these circumstances the embryos of viviparous species will be less vulnerable than the larvae of oviparous ones. A corollary of this idea is that selection might favour viviparity in marginal habitats.

It is dubious practice to select evidence in favour of a hypothesis, and it is certainly not true that all viviparous species occur in marginal or fluctuating habitats. However, a large enough number of instances of viviparity are associated with such areas to suggest that this association may be significant. Two possible examples are the viviparous cave-dwelling animals such as *Proteus* (p. 29), and the reptiles of temperate climates; of the ten most northerly species of snake and lizard in North America and Europe, seven are viviparous.

A situation where a high incidence of viviparity appears in association with *fluctuation* in conditions, rather than with harsh conditions *per se*, is in the marine fauna of the Antarctic. In several groups of Antarctic animals, viviparity and forms of brood care of the young are much more common than in other waters (including the Arctic). In the Holothuria (sea cucumbers), for instance, viviparity and brooding are unusual; but of 38 Antarctic species, at least 15 retain their young in one way or another. In another group of the echinoderms, about 50% of the Antarctic species show brooding or viviparity, compared with about 1% elsewhere. Antarctic waters are characterized by the very seasonal availability of phytoplankton on which, directly or indirectly, all animals depend.

There is, therefore, enough circumstantial evidence to suggest that viviparity may, in some cases, have arisen as an appropriate strategy for survival in fluctuating environments. It would be interesting to compare experimentally the performance of comparable oviparous and viviparous species in controlled environments, both stable and fluctuating.

In considering *why* viviparity should have evolved, the trouble is that all of the evidence is circumstantial, and all of the arguments speculative. Since viviparity takes such diverse forms, and has evolved many times, presumably the advantages which were conferred were equally diverse. There are good reasons to suppose that there is no universal advantage in being viviparous.

Glossary

The principal reproductive hormones of female mammals

FSH: Follicle stimulating hormone. A glycoprotein hormone produced by the anterior pituitary; stimulates growth of ovarian follicles and (with other hormones) may be luteotrophic in some species.

HCG: Human chorionic gonadotrophin. A placental hormone produced by the trophoblast of the human placenta; probably replaces pituitary gonadotrophin in maintaining the corpus luteum.

LH: Luteinizing hormone. Glycoprotein, secreted by the anterior pituitary. It stimulates steroid production by the ovary, ovulation, and the formation of corpora lutea; in some species forms part of the 'luteotrophic complex' of hormones together with prolactin.

Oestrogen: A family of steroids, of which oestradiol-17β is the most important. Produced in the ovary, and also in the placenta in some species; stimulates the LH surge which in turn induces ovulation; prepares and maintains the uterus for pregnancy and parturition; stimulates growth of the mammary glands; induces oestrous behaviour in some species.

Oxytocin: An oligopeptide (for structure, see Fig. 3–3, p. 21) secreted by the posterior pituitary, which stimulates the contraction of smooth muscle. It thus has a major role in expelling milk from the mammary glands, and is also involved in parturition and (possibly) sperm transport in the female genital tract after copulation. Release of oxytocin is evoked by suckling, and by distension of the vagina in parturition or coitus.

Progesterone: Steroid hormone produced mainly by the corpus luteum and, in some species, the placenta. Responsible for maintaining the uterus during pregnancy, and for suppressing premature contraction of uterine or mammary gland smooth muscle.

Prolactin: A protein hormone secreted by the anterior pituitary; involved in preparing the mammary glands for lactation; together with LH forms the 'luteotrophic complex' of hormones in some species.

Prostaglandins: A family of lipids with a wide variety of effects. Prostaglandin $F_2\alpha$ is luteolytic in some species, induces smooth muscle contraction and probably plays an important role in parturition.

Releasing factors: Oligopeptides produced in the hypothalamus, which pass to the pituitary and there control release of the pituitary gonadotrophins. LH and FSH are probably under the control of a single RF. The amount required to have an effect on a rat pituitary may be as little as 5×10^{-9} g. Prolactin is under the control of prolactin inhibiting factor.

Further Reading

Only brief reviews of viviparity as a phenomenon have been published; the most recent is:

AMOROSO, E. C. (1968). The evolution of viviparity. *Proc. roy. Soc. Med.* **61**, 1188–200.

This paper is analysed, and some points amplified, in an Open University Course Unit: *Regulation and the Embryonic Environment*. Course Unit 9 of O.U. Course S321, Physiology of Cells and Organisms (1974).

Other general references are:

BARRINGTON, E. J. W. (1964). *Hormones and Evolution*. English Universities Press, London.

PARKES, A. S., Ed. (1964–5). *Marshall's Physiology of Reproduction*. 3rd edn. Longman, London.

VAN TIENHOVEN, A. (1968). *Reproductive Physiology of Vertebrates*. W. B. Saunders, Philadelphia and London.

Otherwise, information is widely scattered in texts, usually on specific taxonomic groups. The following are useful:

AUSTIN, C. R. and SHORT, R. V. (1972). *Reproduction in Mammals*. Vol. 2: *Embryonic and Fetal development*. Vol. 3: *Hormones in Reproduction*. Vol. 4: *Reproductive Patterns*. Cambridge University Press, London.

BEER, A. E. and BILLINGHAM, R. E. (1974). The embryo as a transplant. *Sci. Amer.*, **230**, 36–46.

BENTLEY, P. J. (1976). *Comparative Vertebrate Endocrinology*. Cambridge University Press, London.

BORRADAILE, L. A., EASTHAM, L. E. S., POTTS, F. A. and SAUNDERS, J. T. (1961). *The Invertebrata*. Cambridge University Press, London.

BREDER, C. M. and ROSEN, D. E. (1966). *Modes of Reproduction in Fishes*. Natural History Press, New York.

BROWNING, H. C. (1973). The evolutionary history of the corpus luteum. *Biol. Reprod.*, **8**, 128–57.

CHIA, F.-S. and ROSTRON, M. A. (1970). Some aspects of the reproductive biology of *Actinia equina* Cnidaria: Anthozoa. *J. Mar. Biol. Assoc. (U.K.)*, **50**, 253–64.

ENGELMANN, F. (1970). *The Physiology of Insect Reproduction*. Pergamon, Oxford.

HOPSON, J. A. (1973). Endothermy, small size, and the origin of mammalian reproduction. *Amer. Nat.*, **107**, 446–52.

NEEDHAM, J. (1950). *Biochemistry and Morphogenesis*. Cambridge University Press, London.

TYNDALE-BISCOE, H. (1973). *Life of Marsupials*. Edward Arnold, London.